Fundamentals of Nuclear Physics

Authored by

Ritesh Kohale
Department of Physics,
Sant Gadge Maharaj Mahavidyalaya,
Hingna-441110,
India

Sanjay J. Dhoble
Department of Physics,
Rashtrasant Tukadoji Maharaj Nagpur University,
Nagpur-440033,
India

Vibha Chopra
Department of Physics and Electronics,
D.A.V. College, Amritsar-143001,
India

Fundamentals of Nuclear Physics

Authors: Ritesh Kohale, Sanjay J. Dhoble & Vibha Chopra

ISBN (Online): 978-981-5049-90-9

ISBN (Print): 978-981-5049-91-6

ISBN (Paperback): 978-981-5049-92-3

Published by Bentham Science Publishers Pte. Ltd. Singapore.

First published in 2023.

Bentham Science Publishers Pte. Ltd.
80 Robinson Road #02-00
Singapore 068898
Singapore
Email: subscriptions@benthamscience.net

CONTENTS

FOREWORD

I welcome the publication of the book 'Fundamentals of Nuclear Physics.' It is an introductory text for scientists, teachers, engineers and students written by Dr. Ritesh Kohale, Dr. Sanjay J. Dhoble and Dr. Vibha Chopra. 'Fundamentals of Nuclear Physics is an ultimate textbook for courses at the undergraduate level in nuclear physics. Moreover, it is a significant basis for scientists and those who initiated working with nuclei, particle physicists, astrophysicists and anyone willing to learn more about novel trends in this field. Its importance on phenomenology and discussions of the theory are covered with the suitable examples that clarify and put on the theoretical formulism, differentiating this book from all other books available. The text is well organized to deliver the fundamentals of content for students with a little mathematical background, providing more distinguished material in each section.

This textbook is competent because it includes the discovery of the neutron and other modern advances, providing readers with widespread coverage of the elementary models of each topic in particle physics for the first time. Physics emphasizes mathematical precision, making the material handy to students with no prior knowledge of elementary nuclear physics. The theory and mathematical expressions are linked together in a very sophisticated manner, helping students understand how key ideas were developed. The content on nuclei, mass defect, packing fraction, particle accelerator and detectors, the discovery of a neutron, a nuclear chain reaction, liquid drop model and the shell model of a nucleus, nuclear fission and fusion, and radioactivity has been completely revised to familiarize the students to what lies beyond. The easily understandable figures and equations with problems inspire students to apply the theory themselves.

I am pleased to authenticate this book as an advanced phase in this ongoing quest, and I hope this book will be a valuable addition to the bare literature on the subject matter.

Dr. K. V. R. Murthy
Department of Applied Physics,
President, Luminescence Society of India (LSI),
M. S. University, Baroda,
India

PREFACE

This textbook clears the cavity between the elementary and the highly progressive volumes that are commonly accessible on the subject. It offers a brief but widespread outline of a number of topics, like fundamental ideas and characteristics of the nucleus, fission and fusion, general relativity, and radioactivity which are otherwise only accessible with much more detail. Providing a general introduction to the underlying concepts allows individuals who read to improve their knowledge of what these two research fields actually encompass. The book uses real-world examples to make the subject more attractive and inspire the use of mathematical formulations. Anticipated essentially for students of scientific disciplines such as physics and chemistry who want to learn about the subject and/or the associated techniques, it is also useful to high school teachers wanting to rejuvenate or modernize their understanding and to fascinated non-experts. This textbook gives an elementary understanding of nuclear and particle physics, offering an overview of theoretical as well as the experimental grounds, providing students with a profound understanding of the ideas about the nucleus, particle detectors, accelerators, radioactivity, and elementary particles, fundamental forces and recent applications of radioactivity. Each chapter provides the fundamental theoretical and experimental knowledge for students to strengthen their concepts regarding nuclear physics.

It is an appropriate textbook for undergraduate courses in nuclear and particle physics as well as more innovative courses; the book includes sophisticated and newly constructed figures as well as thoroughly solved equations that create the interest amongst undergraduate students to renovate the content of their course. It could be a vital textbook for students framing their future study or a profession in the field who needs a concrete understanding of nuclear and particle physics together. It provides a concise, thorough, and accessible treatment of the fundamental aspects of nuclear physics. Reorganized figures, resolved equations, rearranged contents and appendices make it easier to use for entire users. Indeed this book is unique because it makes important connections to other fields such as elementary particle physics and astrophysics. Moreover, its presentation is student-friendly and bridges nuclear physics as an essential part of modern physics with a comprehensive scientific and historical context. We trust that it may be advantageous for graduate students, or more commonly scientists, in various fields. In the first three chapters, we present the "extract," *i.e.*, we give the basic concepts essential to improve the rest. Chapter 1 deals with the introduction to nuclei and basic concepts in nuclear physics. In chapter 2, we describe nuclear fission and fusion. Chapter 3 is dedicated to the nuclear structure and properties of nuclei. Chapter 4 goes a step further. It deals with particle detectors. We shall see that it is conceivable to give a reasonably modest but comprehensive explanation of the major development in particle physics and fundamental dealings made since the late 1960s. In chapters 5 and 6, we turn to the important practical applications of nuclear physics, *i.e.*, particle accelerators and nuclear reactors. In chapters 7 and 8, we intend to understand the origin and applications of radioactivity with some contemporary illustrations of how radioactivity originated and was used, be it in medicine, in the food industry or in engineering. Chapters 9 and 10 are subjected to nuclear astrophysics, stellar structure and evolution and nuclear cosmology and elementary particles. To conclude, we present an introduction to present ideas about nuclear astrophysics in chapter 10.

We want to extend our sincere thanks to all our collogues who constantly provided us with ideas before initiating this project. We are obliged to our co-workers for their irreplaceable help and recommendation throughout the years. We are also grateful to all who directly or indirectly contributed to illuminating discussions on various aspects of nuclear physics.

This book has been kept on track and seen through to completion with the support and encouragement of numerous people, including our well-wishers, our friends, as well as various institutions and laboratories. At the end of this book, it is a pleasant task to express our thanks to all those who contributed in many ways to the success of this study and made it an unforgettable experience for us to write this book.

ACKNOWLEDGEMENTS

Authors express their sincere thanks to Dr. Subhash Chaudhari, Vice Chancellor , Rashtrasant Tukadoji Maharaj Nagpur University, MH, India for his kind co-operation and we owe a great deal of appreciation and gratitude towards Dr. Sanjay Dudhe, Pro-Vice Chancellor, Rashtrasant Tukadoji Maharaj Nagpur University, MH, India for all the help extended by them during the completion of this book. At this moment of accomplishment we would like to express our cordial and honest gratitude to Dr.Arti Moglewar, Principal, Sant Gadge Maharaj Mahavidyalaya, Hingna, MH, India for her guidance, support and constant encouragement. We are also thankful to Dr. Rajesh Kumar, Principal, AV College, Amritsar, India for his valuable advice, constructive criticism and his extensive discussions around this work.

We extend our sincere thanks to all of them who directly and indirectly supported willingly and selflessly for our experiments and all of them who also stayed with us in the process. We are indebted to our many young and dynamic friends for always providing a stimulating and fun filled environment whenever we go into this process.

We can see the good shape of present book because of the enough size of our research group. We remember our communication and coordination with each other, we had interesting and cheerful discussions started from search to research. We enjoyed the same with tea and snacks during the entire process of this book project.

Ritesh Kohale
Department of Physics,
Sant Gadge Maharaj Mahavidyalaya,
India

Sanjay J. Dhoble
Department of Physics,
Rashtrasant Tukadoji Maharaj Nagpur University,
India

&

Vibha Chopra
P.G. Department of Physics and Electronics,
D.A.V. College,
India

CHAPTER 1

Fundamentals of Nuclear Physics

Abstract: The present chapter is an introduction to various scientific and technological fields. It is a beginning step to trail further study in this book. The first chapter specifies the contemporary idea and fundamental understandings of nuclear physics, which is necessary to develop the rest of the studies in this domain. The present chapter deals with an introduction to nuclei, constituents of the nucleus and its properties, mass defects and binding energy, nuclear reactions, the Q-value of nuclear reactions, and the discovery of the neutron and nuclear chain reactions.

Keywords: Chain Reaction, Mass Defect, Neutron, Proton.

1. INTRODUCTION TO NUCLEI

Nuclear physics aims to improve the knowledge of all nuclei and understand astrophysical nucleosynthesis. There are similarities between the electronic structure of atoms and nuclear structure. In nuclei, protons and neutrons are two groups of similar particles [1].

1.1. Constituents of the Nucleus and its Properties

1.1.1. Proton

It is one of the major and significant constituents of the nucleus. Its positive charge and mass are 1.673×10^{-27} kg ≈ 1u.

1.1.2. Neutron

- Zero charge
- Mass 1.675×10^{-27} kg ≈ 1u
- Mass of neutron ≈ mass of proton + mass of an electron

1.1.3. Nucleon

In chemistry and physics, a nucleon is either a proton or a neutron, considered in its role as a component of an atomic nucleus. (*e.g.*, neutron or proton).

Ritesh Kohale, Sanjay J. Dhoble & Vibha Chopra

1.1.4. Nomenclature

- A - Number of nucleons (atomic mass number)
- Z - Number of protons
- N - Number of neutrons
- A = Z + N
- The symbol for the nucleus of chemical element X.

1.1.5. Atomic Mass Unit, (U)

A convenient unit for measuring nuclear mass. Usually, nuclear mass is expressed in terms of a unit known as the *atomic mass unit,* denoted by the letter *u.* One atomic mass unit is defined as one-twelfth the mass of a carbon atom (*i.e.*, the most abundant isotope of Carbon),

viz., ^{12}C, If Mass of ^{12}C = 12 u, then,

$$1\ \text{u} = \frac{1}{12} \text{ x mass of one carbon atom}$$

$$= \frac{1}{12} \text{ x } \frac{12}{6.0 \text{x } 10^{23}}$$

$$1\text{u} = 1.66 \text{ x } 10^{-27} \text{ kg}$$

In terms of the above units,

Mass of a proton, m_p = 1.00727 u

Mass of a neutron, m_n = 1.00866 u

1.1.6. Nuclear Size and Density

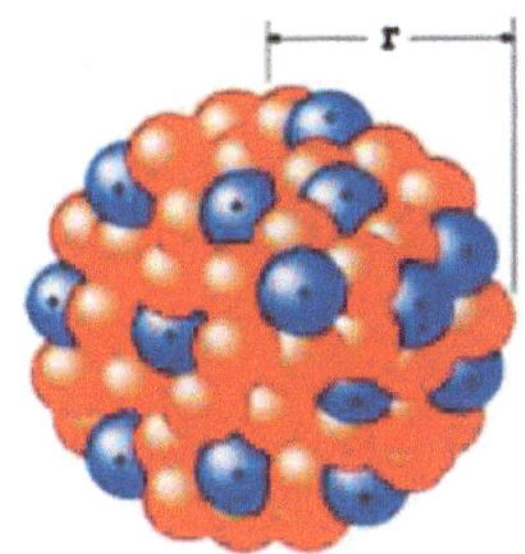

- It has Close-packed structure
- Constant density
- Volume proportional to atomic number (A)
- Since V = 4/3 πr^3, A proportional to r^3
- r is proportional $A^{1/3}$
- $r \approx (1.2 \times 10^{-15}$ m) $A^{1/3} = 1.2$ fm $A^{1/3}$
- Density of neutron star = 100 million tons/cm^3

1.1.7. Nuclei

Nuclei are composed of protons and neutrons. The number of protons is atomic number Z, the number of neutrons is N, and the mass number A is approximately the total number of nucleons (*i.e.*, protons and neutrons, A = (Z + N,)

Therefore the number of neutrons is N = A-Z

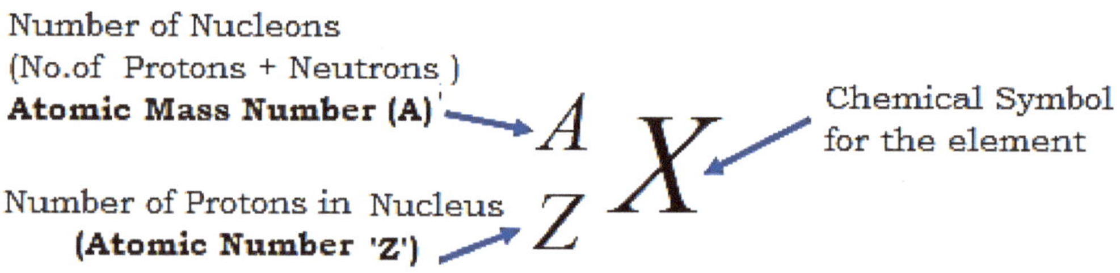

Number of Nucleons
(No.of Protons + Neutrons)
Atomic Mass Number (A) → $^{A}_{Z}X$ ← Chemical Symbol for the element

Number of Protons in Nucleus
(Atomic Number 'Z')

Example: Carbon $^{12}_{6}C$
6 Protons + 6 Neutrons

Hydrogen Nucleus: $^{1}_{1}H$
Helium: $^{4}_{2}He$
Neutron: $^{1}_{0}n$
Electron: $^{0}_{-1}e$
Aluminium: $^{27}_{13}Al \equiv {}^{27}Al$

1.2. Mass Defect and Binding Energy

1.2.1. Mass Defect

The constituents of a nucleus are neutrons and protons, collectively known as nucleons. The mass of a nucleus is always less than the sum of the masses of its constituent nucleons (*i.e.*, protons and neutrons); the difference between the total mass of nucleons and the actual mass of the nucleus is called **the mass defect** denoted by **Δm** and given by [2].

$$\Delta m = [Zm_p + (A-Z)\, m_n] - M$$

Where,

Δm = Mass Defect

Z = no of protons inside the nucleus

(A-Z) = N = no of neutrons

m_p = mass of a single proton

m_n = mass of a single neutron

M = total actual mass of a nucleus

The "missing" mass appears in the form of energy holding the nucleus together,

This energy equivalent to mass defect is called the **binding energy** of the nucleus [3, 4].

Example: The atomic mass of helium (^4He) atom is 4.002602 u. Determine its mass defect.

Solution:

Mass of 2 protons: (2 x 1.007276 u) = 2.014552 u

Mass of 2 neutrons: (2 x 1.008665 u) = 2.017330 u

Mass of 2 electrons: (2 x 0.000548 u) = 0.001097 u

Total combined mass: = 4.032979 u ≠ 4.002602 u

The actual atomic mass of helium (^4He) atom is 4.002602 u, which is not equal to the combined mass of its constituent, *i.e.*4.032979 u.

Mass Defect Δm is given by equation 1.1,

$$\Delta m = [Zm_p + (A\text{-}Z)\, m_n] - M \quad \textbf{(1.1)}$$

$\Delta m = [(2 \text{ x } 1.007276) + (2 \text{ x } 1.008665) + (2 \text{ x } 0.0005486)] \text{ u} - 4.002602 \text{ u}$

$\Delta m = 4.032979 \text{ u} - 4.002602 \text{ u}$

$\Delta m = 0.030366 \text{ u}$

This difference between the mass of an atom and the sum of the masses of its protons, neutrons, and electrons is 0.030366 u, which is *less* than the combined mass called the mass defect (Fig. **1.1**).

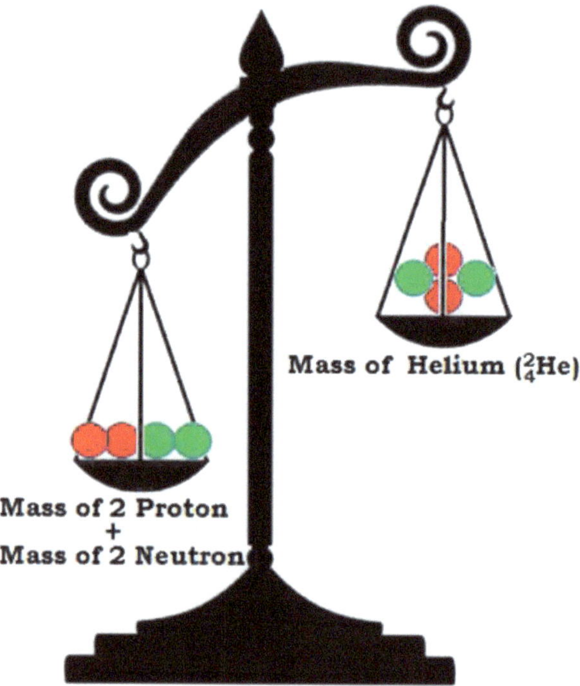

Fig. (1.1). The schematic shows the actual atomic mass of the Helium atom and the combined mass of its constituents.

1.2.2. Binding Energy

As the mass of a nucleus is always less than the sum of the masses of its constituent nucleons, the difference between the two is called **the mass defect.** This loss of mass is equivalent to the energy given by Einstein's Mass-Energy correlation: $E=\Delta m.c^2$ and is called the **binding energy** of the nucleus.

Thus, the binding energy of a nucleus can be defined as the total amount of energy released when nucleons combine to form stable nuclei or the amount of external energy required to separate the nucleus into its constituent nucleons.

From special relativity, adding energy increases mass which is a measure of binding energy and can be given as:

$$B.E. + mc^2 = m_1c^2 + m_2c^2 + ... = \sum m_i c^2$$

$$B.E. = \left\{\sum m_i - m\right\}c^2 \qquad \textbf{(1.2)}$$

$$B.E. = \Delta mc^2$$

Δm is a mass defect, and binding energy in terms of atomic number Z and atomic mass number A is,

From equations (1.1) and (1.2)

$$\text{B.E.} = [\text{Zm}_\text{p} + (\text{A-Z})\ \text{m}_\text{n} - \text{M}] \text{ x } \text{c}^2 \qquad \textbf{(1.3)}$$

Where, M is the mass of the combined nucleus, m_p is the mass of the proton, m_n is the mass of the neutron, and c is the velocity of light.

Average Binding Energy: It is defined as binding energy (B.E.) per nucleon (A) and given by,

$$\frac{\text{B.E.}}{\text{A}} = \frac{\Delta\text{mc}^2}{\text{A}} = \frac{[\text{Zm}_\text{p} + (\text{A}-\text{Z})\text{m}_\text{n} - \text{M}]\times\text{c}^2}{\text{A}} \qquad \textbf{(1.4)}$$

$$\text{B. E.} = \text{c}^2 \times \left[\text{m}_\text{n} - \frac{\text{Z}}{\text{A}}(\text{m}_\text{n} - \text{m}_\text{p}) - \frac{\text{M}}{\text{A}}\right] \qquad \textbf{(1.5)}$$

The term $\left[\text{m}_\text{n} - \frac{\text{Z}}{\text{A}}(\text{m}_\text{n} - \text{m}_\text{p}) - \frac{\text{M}}{\text{A}}\right]$ in equation (1.5) is known as the ***packing fraction***.

i.e., **Binding Energy (B.E.) per nucleon = c^2×Packing Fraction**

The nuclei with the greatest binding energy per nucleon are the most stable. For a given number of nucleons, if the total mass defect increases, then BE per nucleon also increases, and energy can be released. Nuclear binding energy is more or less independent of the nuclei's size and is roughly about 8.5 MeV/nucleon. Most stable nuclides, from the lightest to the most massive, have binding energies of 7 to 9 MeV/nucleon. It may be noted that the nucleus Fe is the most tightly bound nucleus with a binding energy of about 8.8 MeV per nucleon. This is why the iron group of nuclei is the most stable. A graph of binding energy per nucleon as a function of mass number A is shown in Fig. (**1.2**) below [5].

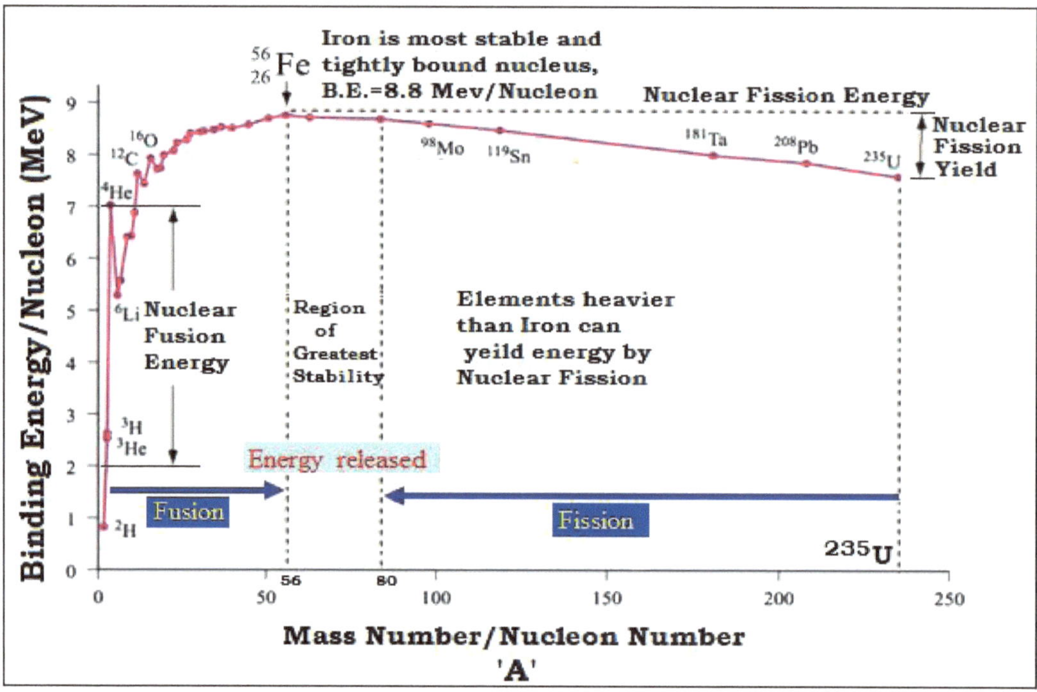

Fig. (1.2). Binding energy per nucleon curve.

1. From Fig. (**1.2**), the maximum value of B.E.$_{ave}$ is for ^{56}Fe. The large binding energy suggests that the nucleons in a nucleus are very tightly bound.

2. **For 0 ≤ A ≤ 30,**

The binding energy per nucleon B.E.$_{ave}$ increases as the mass number 'A' increases. At some particular values of A, the value of B.E.$_{ave}$ is apparently larger than neighboring *A*'s.

3. **For 30 ≤ A ≤ 240,**

The value of B.E.$_{ave}$ varies slowly and is around 7.5 ~ 8.5 MeV.

4. **For ≈ 56 (*viz.* 56_{Fe})**

The value of B.E.$_{ave}$ reaches its maximum value.

5. **For A>60**

The value of B.E.$_{ave}$ decreases as the mass number ‘A’ increases. This is due to the Coulomb repulsion.

6. The binding energy per nucleon for light nuclei such as ^{1}H, ^{2}H and ^{3}H is approximately low, but it is about 8 MeV for all other nuclei.

1.3. Nuclear Reaction

A nuclear reaction is a process in which the target nucleus of an atom is bombarded by fast-moving particles that split apart or are joined with the nucleus of another atom.

In general, the nuclear reaction can be written as:

$$\boldsymbol{a + X \longrightarrow Y + b + Q} \tag{1.6}$$

Where,

a- Is incident particle(Projectile)

X - Is the target nucleus

Y - Is the product nucleus (as X changes into Y)

b - Is the product particle

Q = energy absorbed/released in the nuclear reaction.

If Q is positive, energy is released and

Q is negative energy is absorbed

The above nuclear reaction can be written as: *X (a, b) Y*

An example of a nuclear reaction is firing α-particles at beryllium:

$$\overset{A}{{}^{4}_{2}\alpha} + \overset{B}{{}^{9}_{4}Be} \rightarrow \overset{C}{{}^{12}_{6}C} + \overset{D}{{}^{1}_{0}n} + Q\ (\text{Energy})$$

Projectile Target Products

4 + 9 = 12 + 1
2 + 4 = 6 + 0

And this can be equivalently written as: ${}^{9}_{4}\mathrm{Be}(\alpha, \mathrm{n})\,{}^{12}_{6}\mathrm{C}$

1.3.1. Q-value of Nuclear Reaction

The Q-value of a nuclear reaction is defined as the total energy released or absorbed during the nuclear reaction.

It is equivalent to the total change in the system's kinetic energy (KE). Depending upon the type of reaction, the Q-value may be positive or negative.

In a generic reaction:

$$A + B \rightarrow C + D$$

$$m_A \quad m_B \quad m_C \quad m_D \Rightarrow \text{Masses}$$

$$K_A \quad K_B \quad K_C \quad K_D \Rightarrow \text{Kinetic Energy}$$

Hence according to the law of conservation of energy,

$$m_A c^2 + K_A + m_B c^2 + K_B = m_c c^2 + K_C + m_D c^2 + K_D$$

From the above reaction, we can calculate the Q-value or the energy released or absorbed during the nuclear reaction as shown below:

$$Q = K_{final} - K_{initial} = (K_C + K_D) - (K_A + K_D) \quad \textbf{(1.7)}$$

$$Q = (K_C + K_D) - (K_A + K_D) \Rightarrow (m_A + m_B)c^2 - (m_C + m_D)c^2 \quad \textbf{(1.8)}$$

Hence,

$$Q = (m_A + m_B)c^2 - (m_c + m_D)c^2 \quad \textbf{(1.9)}$$

$$Q = [(m_A + m_B) - (m_C + m_D)]\, c^2 \tag{1.10}$$

Hence, the Q-value of nuclear reaction is defined as the difference between the kinetic energies of the products and that of the incident particle.

1. If $(\mathbf{m_A} + \mathbf{m_B}) > (\mathbf{m_C} + \mathbf{m_D})$,

Then Q-value is positive, and the reaction is exothermic.

i.e., Energy is released in the reaction.

2. If $(\mathbf{m_A} + \mathbf{m_B}) < (\mathbf{m_C} + \mathbf{m_D})$,

Then the Q-value is negative, and the reaction is endothermic.

i.e., Energy is absorbed in the reaction.

3. If $(\mathbf{m_A} + \mathbf{m_B}) = (\mathbf{m_C} + \mathbf{m_D})$,

Then the Q-value is zero, and the reaction is a kind of elastic collision.

i.e., Energy is neither released nor absorbed in the reaction.

1.4. Discovery of Neutron

One of the famous experiments performed by Rutherford was the bombarding of lighter elements by α - particles, *e.g.*, When nitrogen was bombarded by ∝ - particles, the following reaction occurred.

$$^{4}_{2}He + {}^{14}_{7}H \rightarrow {}^{1}_{1}H + {}^{17}_{8}O$$

This was one of the first man-made nuclear reactions. Rutherford and his associate James Chadwick performed experiments with many light nuclei where bombardment of light nuclei by cc-particles resulted in the ejection of protons. However, it was soon found that some light elements like boron and beryllium did not give out protons under α-particle bombardment. In 1930, W-Bothe and H. Becker found that when Beryllium or Boron was bombarded with ∝-particles (of Polonium), highly penetrating radiation was emitted. This radiation was assumed to be γ-rays since γ-rays have a high penetrating power. Further, in 1932), Irene-Curie-Joloit (daughter of Madame Curie) and her husband, F. Joliot, found that these penetrating radiations could knock out energetic protons from

Paraffin wax. They thought these penetrating radiations in the town of γ-rays knocked out protons in the same manner as X-rays knocked out electrons in Compton Effect. Further calculations show that to knock out protons of energy 4.7 MeV from paraffin wax, the γ-rays must have an energy of about 47 MeV. This much energy is unusually high for γ-rays, and so this situation is confusing. Soon afterward, in 1932, Chadwick repeated the experiment with paraffin wax, helium and nitrogen. Chadwick resolved the anomaly of energy of penetrating radiations by assuming that the unknown radiation produced by the impact of α - particles with $_4Be^9$ nuclei was not γ-ray but a new type of particle of the same mass as the proton and was electrically neutral [6]. This new particle was named 'neutron.' The reaction proposed by Chadwick could be written as:

$$ {}^{9}_{4}Be + {}^{4}_{2}He \rightarrow \frac{13}{6}C \rightarrow {}^{12}_{6}C + {}^{1}_{o} $$

Since a neutron has no charge, it can have great penetrating power. Chadwick was awarded Nobel Prize in 1935 for this great discovery.

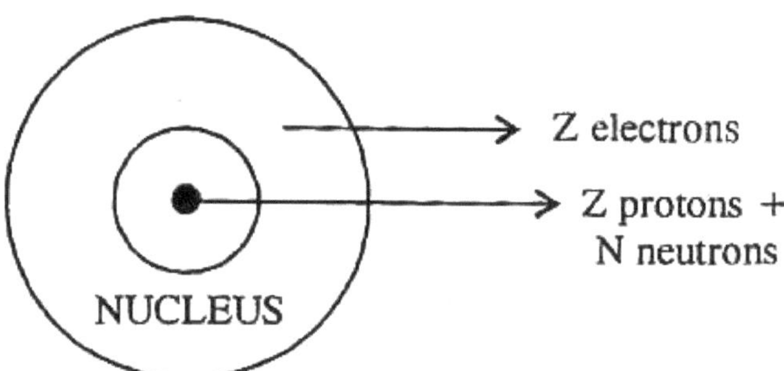

In 1930, Rutherford, in his gold foil experiment, bombarded a beam of alpha particles on an ultrathin gold foil, which explained the overall electrical neutrality of an atom and gave a clear picture of the structure of an atom (which consists of protons (inside the nucleus) and the same number of electrons outside of the nucleus).

But scientists soon realized that Rutherford's atomic model was incomplete. Various experiments showed that the nucleus's mass is approximately twice the number of protons. Then what is the origin of this additional mass? Rutherford postulated the existence of some neutral particles having mass similar to protons, but there was no direct experimental evidence.

Several theories and experimental observations eventually led to the discovery of neutrons [7]. We can summarize some of the scientific observations behind the discovery of neutrons.

- In 1930, W. Bothe and H. Becker found electrically neutral radiation when they bombarded beryllium with an alpha particle. They thought it was photons with high energy (gamma rays).
- In 1932, Irene and Frederic Joliot-Curie showed that this ray could eject protons when it hits paraffin or H-containing compounds.
- The question arose about how massless photons could eject protons 1836 times heavier than electrons. So the ejected rays in the bombardment of beryllium with alpha particles cannot be photons.
- In 1932, *James Chadwick* performed the same experiments as Irene and Frederic Joliot-Curie but used many different targets of bombardment besides paraffin (Fig **1.3**).
- By analysing the energies of different targets after bombardment, he discovered the existence of a new particle that is charged less and has a similar mass to a proton [8]. This particle is called a ***neutron.***

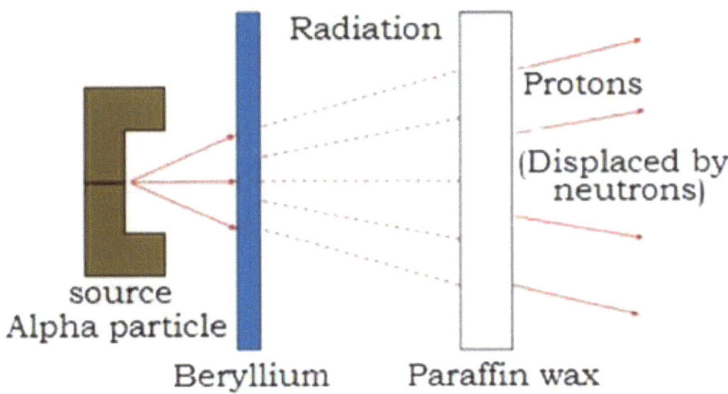

Fig. (1.3). Schematic diagram for the experiment that led to the discovery of neutrons by Chadwick.

Example: Beryllium undergoes the following reaction when bombarded with an alpha particle.

$$^{9}_{4}\mathrm{Be} + {}^{4}_{2}\mathrm{He}\ (\alpha\ \mathrm{Particle}) \longrightarrow [{}^{13}_{6}\mathrm{C}] \longrightarrow {}^{12}_{6}\mathrm{C} + {}^{1}_{0}\mathrm{n}$$

1.5. Nuclear Chain Reaction

When fission takes place for heavy atoms, it releases a number of multiplying neutrons which can be absorbed by other heavy atoms to induce further fissions; this is called a chain reaction.

Notice that in each of the reactions, neutrons are produced. If each neutron releases two more neutrons from fission, then the number of neutrons doubles in each fission generation (Fig **1.4**); in that case, in 10 generations, there are 1,024 fissions, and in 80 generations, about 6 x 10^{23} (a mole) fissions [9].

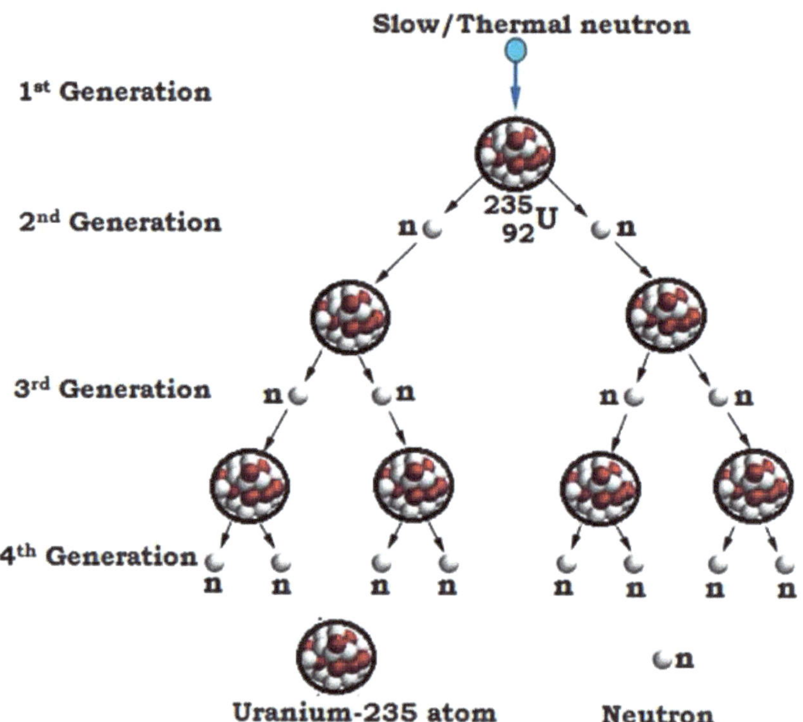

Fig. (1.4). Nuclear chain reaction.

Consider nuclear chain reactions for ^{235}U, where two neutrons are produced. The first neutron causes one uranium nucleus to split into two fragments producing two neutrons, which, in turn, can cause two more uranium to fission, producing four fragments and four new neutrons. The chain reaction will continue till the total ^{235}U fuel is spent. Thus in a nuclear chain reaction, an incredible amount of energy is released in a very short time. Fig. (**1.5**) shows schematics of nuclear chain reaction for uranium (^{235}U).

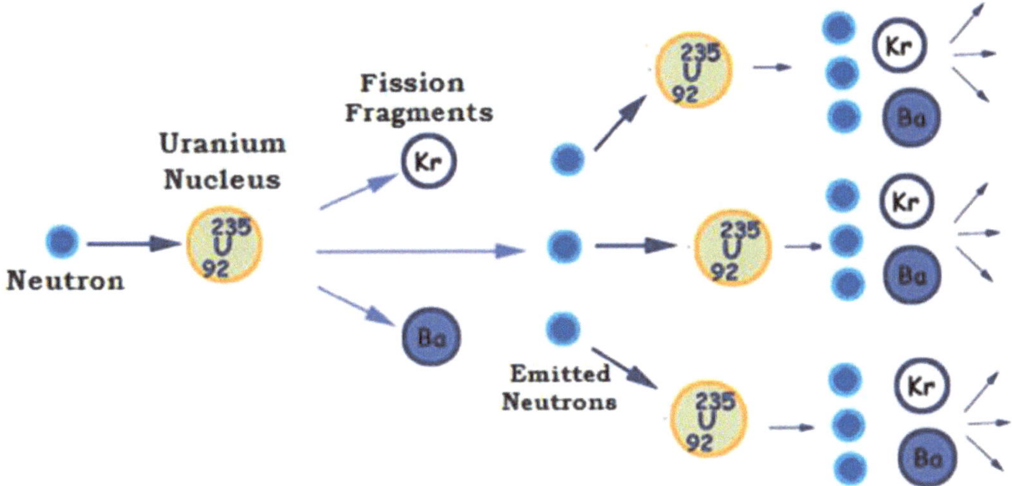

Fig. (1.5). Nuclear Chain Reaction in Uranium 235.

This is roughly what happens in an atom bomb. The run-away chain reaction needs to be controlled if one is to use it for power generation. This is done by removing excess neutrons so the chain reaction can proceed slowly until all the fuel is burnt. Removing excess neutrons is achieved by inserting control rods that contain neutron-absorbing material. Cadmium and boron rods are usually used for such purposes [10].

1.5.1. Types of Chain Reaction

In a chain reaction, the process may be controlled by neutron-absorbing materials (as in nuclear power reactors) or uncontrolled (as in nuclear weapons). There are two basic types of chain reactions.

1.5.1.1. Controlled Chain Reaction

If the chain reaction rate can be sustained at a constant level and the rate of neutron production equals the rate of neutron loss, then the reaction is called a controlled or critical chain reaction.

1.5.1.2. Uncontrolled Chain Reaction

If the chain reaction rate rises, the rate of neutron production is greater than the rate of neutron loss; the reaction is called an uncontrolled or supercritical chain reaction.

REFERENCES

[1] Particle Data Group, *Eur. Phys. J. C,* vol. 15, p. 1, 2000.

[2] *E. Segre from X rays to Quarks.* Freeman: San Francisco, 1980.

[3] A. Pais, *Inward Bound.* Oxford University Press: Oxford, 1986.

[4] R. Hofstadter, "Nuclear and Nucleon Scattering of High-Energy Electrons", *Annu. Rev. Nucl. Sci.,* vol. 7, no. 1, pp. 231-316, 1957. [http://dx.doi.org/10.1146/annurev.ns.07.120157.001311]

[5] G. Audi, and A.H. Wapstra, "The 1993 atomic mass evaluation", *Nucl. Phys. A.,* vol. 565, no. 1, pp. 1-65, 1993. [http://dx.doi.org/10.1016/0375-9474(93)90024-R]

[6] K.S. Quisenberry, T.T. Scolman, and A.O. Nier, "Atomic Masses of H 1, D 2, C 12, and S 32", *Phys. Rev.,* vol. 102, no. 4, pp. 1071-1075, 1956. [http://dx.doi.org/10.1103/PhysRev.102.1071]

[7] R.B. Firestone, V.S. Shirley, C.M. Baglin, S.Y.F. Chu, and J. Zipkin, *Table of Isotopes.* Wiley, 1996.

[8] D.H. Clark, and F.R. Stephenson, *The Historical Supernovae.* Pergamon Press: Oxford, 1977.

[9] P.G. Hansen, A.S. Jensen, and B. Jonson, "Nuclear Halos", *Annu. Rev. Nucl. Part. Sci.,* vol. 45, no. 1, pp. 591-634, 1995. [http://dx.doi.org/10.1146/annurev.ns.45.120195.003111]

[10] H. Hintenberger, W. Herr, and H. Voshage, "Radiogenic Osmium from Rhenium-Containing Molybdenite", *Phys. Rev.,* vol. 95, no. 6, pp. 1690-1691, 1954. [http://dx.doi.org/10.1103/PhysRev.95.1690]

CHAPTER 2

Nuclear Fission and Fusion

Abstract: The present chapter deals with the analysis and relationship of significant features of theoretical nuclear physics. It is perhaps the most widely adopted chapter on the subject. The authors' line of understanding is subjected to "the theoretical perceptions, approaches, and deliberations formulated to infer the investigational matter and spread our aptitude to calculate and govern nuclear occurrences." The present chapter elaborates on the features of conjectural nuclear physics. Its attention is classified agreeing to occurrences concerning nuclear fission, transition state (saddle point) and scission point, photo–fission, fissile materials and fertile materials, moderation and thermalization of the neutron, neutron transport in the matter, nuclear fusion and basic reaction for energy generation in the sun by fusion.

Keywords: Fission, Fusion, Neutron Transport, Photo–fission.

1. INTRODUCTION

Energy from the nucleus draws attention to the two central approaches to generating energy from the nucleus: fission and fusion. In the present approach, the eminence of existing and upcoming reactors advanced security provisions and the eco-friendly effect of electrical energy generation from nuclear fission. The sections in the chapter proceeding with nuclear fusion address both inertial and magnetic confinement fusion.

The significant aspect of fission is photo-fission was discovered in 1940 by a small team of engineers and scientists functioning the Westinghouse Atom Smasher at the company's Research Laboratories in Forest Hills, Pennsylvania [1]. They used a 5 MeV proton beam to bombard fluorine and produce high-energy photons, which were then exposed to samples of uranium and thorium [2]. In the low tens of MeV, Gamma radiation of modest energies can induce fission in conventionally fissile essentials such as the actinides thorium, uranium [3], plutonium, and neptunium [4]. Experiments have been conducted with much higher energy gamma rays, finding that the photo fission cross-section varies little within ranges in the low GeV range.

Ritesh Kohale, Sanjay J. Dhoble & Vibha Chopra

Fission is a system of nuclear transformation because the consequential fragments (or daughter atoms) are not the identical constituent as the unique parent atom. The two (or more) nuclei formed are most regularly of similar but somewhat different sizes, characteristically with a mass ratio of products of about 3 to 2 for common fissile isotopes. Most fissions are binary (generating two charged fragments), but sometimes (2 to 4 times per 1000 events), three positively charged fragments are formed in a ternary separation. The tiniest of these fragments in ternary developments sorts in size from a proton to an argon nucleus.

In 1920, Arthur Eddington recommended that hydrogen-helium fusion could be the principal basis of cosmological energy. Quantum channelling was revealed by Friedrich Hund in 1929, and soon after, Robert Atkinson and Fritz Houtermans used the measured masses of light components to show that large extents of energy could be released by combining small nuclei. Building on the early experiments in simulated nuclear transformation by Patrick Blackett, a laboratory with an emerging and effective fusion inside a fusion container or reactor has continued since the Berkeley cyclotron get up in the 1940s, but the expertise and equipment for fusion are still in their improvement stage. The objective of this section in the book is to recognize by what means definite arrangements of N neutrons and Z protons produce certain states. The ample nuclear classes encompass additional neutrons or protons; thus, they are β-unstable. Many dense nuclei degenerate by giving out of α-particle or by additional arrangements of allowed fission into lighter components. The further purpose of the present chapter is to recognize why definite nuclei are steady compared to these degenerations and anything that regulates the leading decay approaches of unstable cores of nuclei.

1.1. Nuclear Fission Basics

Nuclear fission is the process of splitting the nucleus of a heavy atom (target nucleus) into two or more lighter atoms (fission products) when a neutron bombards the heavy atom.

Fission releases a large amount of energy along with two or more neutrons due to the sum of the masses of the fission products being less than the original mass of the heavy atom.

Otto-Hann and Stresemann bombarded ^{235}U with thermal neutrons and observed that the Uranium nucleus splits into two nuclei, ^{141}Ba and ^{92} Kr, with the emission of 3 neutrons, and hence, a great extent of energy is liberated. The basic nuclear fission of ^{235}U can be given as,

$$^{235}U + {}^{1}n \rightarrow {}^{236}U \rightarrow {}^{141}Ba + {}^{92}Kr + 3\ {}^{1}n + \text{energy}$$

The products formed during nuclear fission are called fission fragments. ^{141}Ba and ^{92}Kr are the fission fragments from the ^{235}U nucleus, see Fig. (**2.1**).

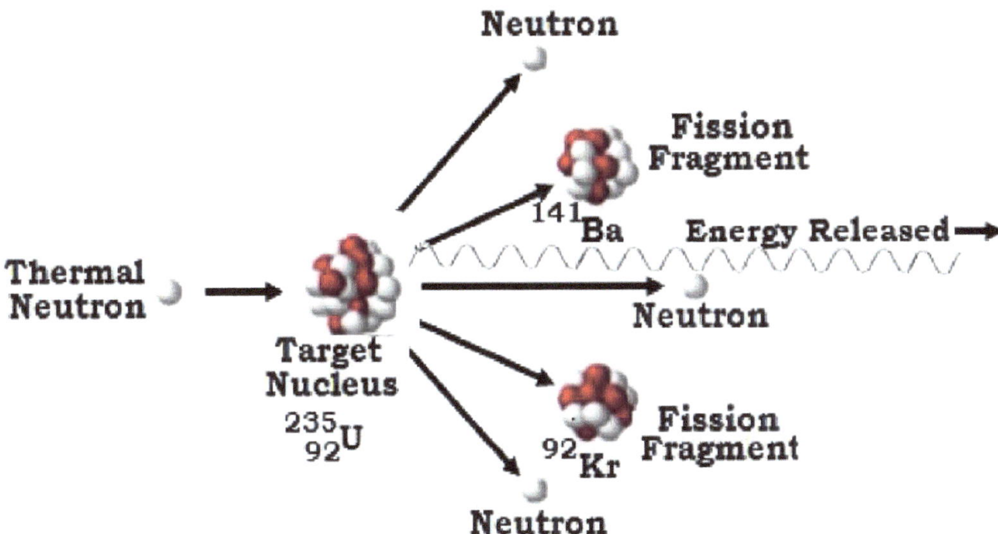

Fig. (2.1). Nuclear Fission Mechanism.

1.1.1. Nuclear Fission Energy

Nuclear energy is taken out by exothermic nuclear reactions; fission and fusion are the two main categories. Obviously, we ought to evoke that radioactivity discharges an enormous amount of energy in this particular situation, as Henri Becquerel and Pierre Curie first expected. For example, the decay of ^{238}U,

$$^{238}U \rightarrow {}^{234}Th\ \alpha$$

$$t_{1/2} = 4.468 \times 10^{9}\ \text{Yr}$$

$$Q_{\alpha} = 4.262\ \text{MeV}$$

Releases the amount of power,

$$P = \frac{N_A}{238}\frac{Q_{\alpha} I_n 2}{t_{1/2}} = 8 \times 10^{-9} Wg^{-1}$$

This small extent of power is enlarged in nuclear reactors up to the tenth times of extent from concluding neutron-prompted fission of ^{235}U. It enhances the Q value from 4.668 MeV to ~ 200 MeV and reduces the actual lifespan to a small figure of months. Nuclear Fission and fusion are the prominent reactions that produce enormous amounts of nuclear energy.

Spontaneous fission falls under the potential barrier problem, as shown in Fig. (**2.2**). The solid line resembles the nature of the potential in the parent nucleus. The activation energy presented in Fig. (**2.2**) decides the possibility of spontaneous fission. As in the fission reaction, the fragments are enormous, and the probability for this to happen is insignificant, so the nucleus may possibly tunnel through the barrier in the fission reaction. Moreover, the activation energy for heavy nuclei is about 6 MeV, but it vanishes for precisely heavy nuclei. For such nuclei, the potential shape relates nearer to the dashed line, and the small amount of deformation will encourage the fission reaction.

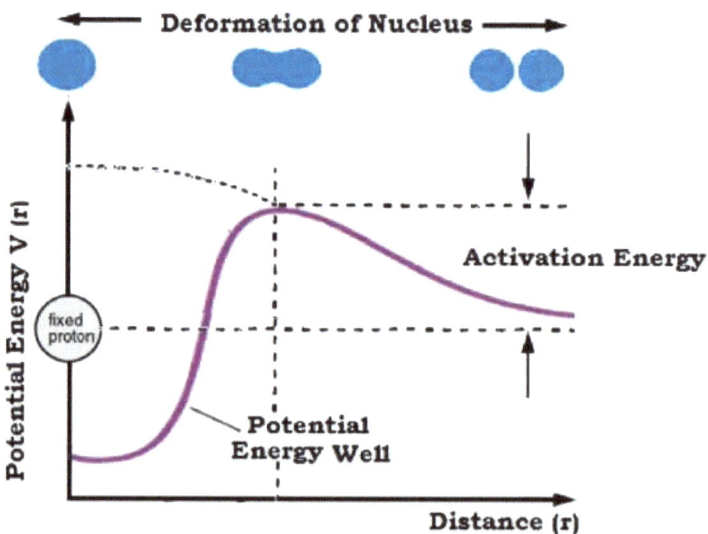

Fig. (2.2). Potential energy during deformation of the nucleus.

Fig. (**2.2**) is intended for slight distortions of the nucleus; the energy increases with increasing deformations due to the increase in the exterior region. It can also be demonstrated that when twofold fragments are disjointed, the energy decreases with expanding separation since the Coulomb energy also decreases. An energy obstruction (activation energy E_A) must be crossed over for fission to happen.

An alternative prospect for fission is to supply the energy required to overcome the barrier by moving neutrons. Because of the lack of a Coulomb force, a neutron can get adjacent to the nucleus and be arrested by a strong nuclear pull. The parent nucleus may then be excited to a state directly above the fission barrier and consequently split up. This progression is a pattern of induced fission [2].

1.1.2. Fission Products

In the fission process, large numbers of particles are emitted due to fission fragments. Depending upon the nature of reaction kinetics, they can be classified as follows:

i. "Rapid" particles, typically neutrons released in the splitting as fission progression takes place and photons released by the principal fission fragments formed in particularly excited conditions of the nucleus.
ii. "Deferred" particles typically $\bar{e}$, e, $\bar{\nu}$ and γ radiations are released in the β^- -decays with the principal fission components and their daughters.
iii. Two fission fragments β^- that are not stable.

The maximum amount of energy emitted is enclosed in the preliminary kinematic energies of the dual fission elements. The kinematic energy (ν) of every single dense (substantial) fragment while deformation takes place is about 75 MeV, with preliminary velocities of approximately 10^7 ms^{-1}. Assuming their enormous masses, their assortments are insignificant ~ 10^{-6} m. The stopping progression converts the kinetic energy to thermal energy.

For an assumed fissionable nucleus, we appeal to the steady amount of free neutrons liberated, κ the number of photons and μ the number of β-decays. Therefore the total energy equilibrium of a fission reaction can be estimated as,

$$A \rightarrow B + C + \nu n + \mu\, e^- + \mu\, \bar{\nu}\, e + \kappa\, \gamma.$$

On average, the numerous constituents yield the subsequent energies as tabulated in Table **2.1**.

Table 2.1. Average energies yields of constituents in the fission process.

S.N.	Energy of Constituents in Fission	MeV
1	Kinetic energy of fragments	165 ± 5
2	Energy of prompt photons	7 ± 1
3	Kinetic energy of neutrons	5 ± 0.5
4	Energy of β decay electrons	7 ± 1

S.N.	Energy of Constituents in Fission	MeV
5	Energy of β decay antineutrinos	10
6	Energy of γ decay photons	6 ± 1
	Total	**200 ± 6**

It is estimated that amongst ~ 200 MeV, only 190 MeV are convenient for utilization; subsequently, the neutrinos emit and do not shatter the medium. The neutrons liberated will continue chain reactions in induced fission mechanisms. The average number of produced neutrons for the fission of 235U prompted by a neutron is ν = 2.47.

The energy distribution of these neutrons peaked around En = 0.7 MeV. The average energy is < En >~ 2MeV.

Statistically, the fission fragments may have dissimilar masses. This is called *asymmetric fission*. Consequently, in neutron prompted splitting of ^{235}U in fission, one can notice:

i. A fission fragment in the "group " $A \sim 95$, $Z \sim 36$ (Br, Kr, Sr, Zr)
ii. A fission fragment in the "group" $A \sim 140$, $Z \sim 54$ (I, Xe, Ba).

A given nucleus can fission in numerous behaviours. Such as for ^{236}U, one probability is:

$$^{236}_{92}U \rightarrow {}^{137}_{53}I + {}^{96}_{39}Y + 3n$$

$$^{137}I \rightarrow {}^{137}Xe\,\overline{\nu e}e^-, \qquad t_{1/2} = 24.5\text{ s}$$

$$^{137}Xe \rightarrow {}^{137}Cs\,\overline{\nu e}e^-, \qquad t_{1/2} = 3.818\text{ m}$$

$$^{137}Cs \rightarrow {}^{137}Ba\,\overline{\nu e}e^-, \qquad t_{1/2} = 30.07\text{ Yr}$$

$$^{96}Y \rightarrow {}^{96}Zr\,\overline{\nu e}e^-, \qquad t_{1/2} = 5.34\text{ s},$$

Or in common, the above reactions can be generalized as,

$$^{236}_{92}U \rightarrow {}^{137}_{56}Ba + {}^{96}_{40}Zr + 3n + 4e^- + 4\nu e$$

The initial fission fragments ^{137}I and ^{96}Y are transmuted over a series of β-decays to the β-stable ^{137}Ba and ^{96}Zr. Note the long half-life of ^{137}Cs categorize productively and steadily to form a complete fuel cycle of a nuclear reactor. It can be considered a versatile illustration of emitting radiation left-over from a fission container or nuclear reactor. Obviously, the equation mentioned above is one of the numerous conceivable fission methods. It is essential to deliberate the problem statistically [3].

The scattering or disintegration for ^{235}U is controlled by unstable separation into a light nucleus ($A \sim 95$) and a heavier nucleus ($A \sim 140$) during fission, realistically near the magic neutron numbers $N = 50$ and $N = 82$ (See Fig. **2.3**).

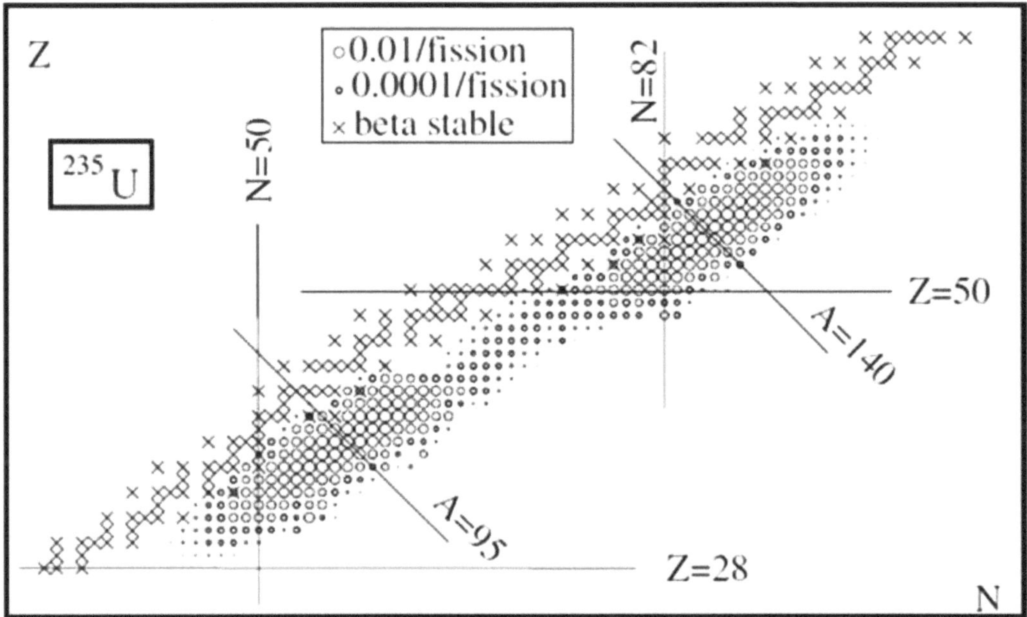

Fig. (2.3). The scattering of fission fragments for neutron prompted the fission of ^{235}U.

The distribution for ^{252}Cf is extensive; on the other hand, statically, it is controlled by unbalanced fission. Because of the enormous neutron surplus in nuclei with $A > 230$, fission remains nearly beneath the line of β-stability and consequently decays by β-emission (See Fig. **2.4**).

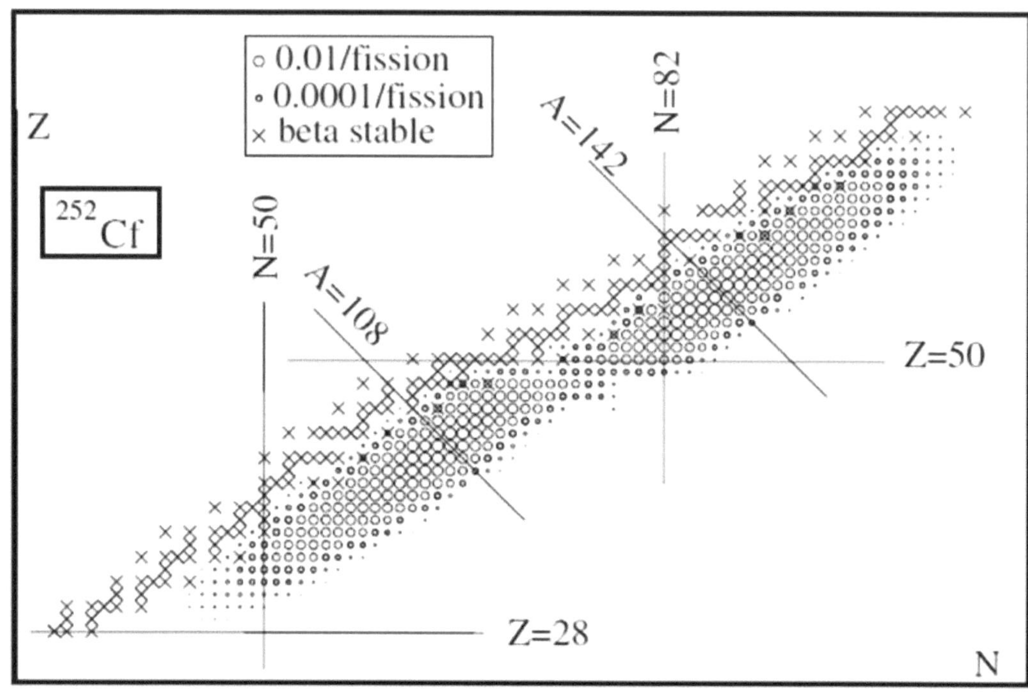

Fig. (2.4). The scattering of fission fragments for unprompted fission of ^{252}Cf.

1.1.3. Fission Barrier

The fission barrier is the activation energy essential for a nucleus of an atom to go through fission. This barrier may also be well-defined as the smallest amount of energy necessary to deform the nucleus to the point where it is permanently committed to the fission process. The energy to overcome this barrier can originate from either neutron bombardment of the nucleus, where the complementary energy from the neutron transports the nucleus to an excited state and experiences deformation or concluded with spontaneous fission, where the nucleus is already deformed and in an excited state.

It is important to note that the determination to recognize the fission process is stagnant and ongoing action and has been a very challenging problem to resolve since its innovation by Otto Hahn and Fritz Strassmann in 1938. Although nuclear physicists appreciate numerous characteristics of the fission process, there is currently no incorporating speculative outline that gives an acceptable explanation of the straightforward clarifications.

1.1.3.1. Transition State (Saddle Point) and Scission Point

The fission process can be assumed when a nucleus with specific equilibrium distortion absorbs energy through neutron capture, turn out to be excited and deforms to a specific shape and size, known as the 'transition state' or 'saddle point' (See Fig. **2.5**) configuration. As the nucleus deforms, the nuclear Coulomb energy reduces even though the nuclear surface energy rises. At the saddle point, the rate of change of the Coulomb energy is in equilibrium with the rate of change of the nuclear surface energy. The materialization and subsequent decay of this conversion state nucleus is the rate-defining phase in the fission process and resembles the channel over an activation energy barrier to the fission reaction. When this phenomenon occurs, the neckline amongst the emerging fragments fades, and the nucleus distributes into two fragments. The point at which this arises is called the 'scission point' (See Fig. **2.5**).

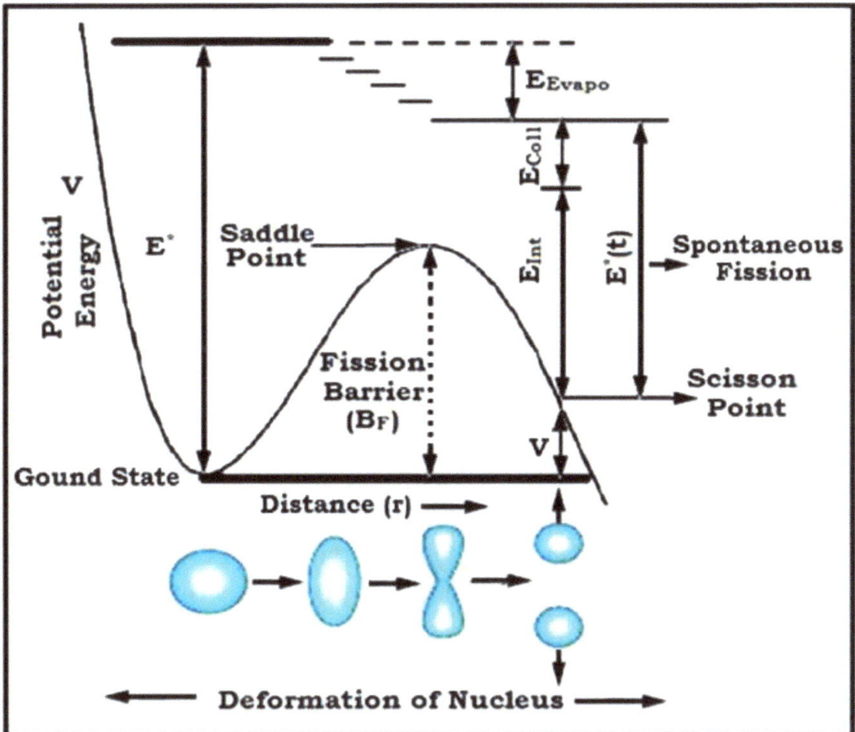

Fig. (2.5). The Saddle point and Scission point during deformation of the nucleus.

Suppose we undertake the potential energy of the dynamical arrangement as a function of distance *r* amongst the two fission fragments designated in Fig. (**2.5**). In the beginning, the two fragments are combined in an essentially spherical nucleus. As distance r rises, the nucleus turns out to be distorted; its outer size increases related to its novel nature. Consequently, this distortion affects the surface tension energy, increasing accordingly. Alternatively, the increase in the distance $A - B$ resembles a reduction of the *Coulomb repulsion* energy *(Coulomb potential, See* Fig. (**2.5**) between A and B. There is a power struggle between the Coulomb repulsion and the nuclear forces. Sometimes, as r changes between r_0 (the original form of the nucleus) and infinity (detached fragments A and B), the potential energy of the system has the utmost value. Moreover, there is a potential barrier, entitled the *fission barrier,* that is essential to cross over for the fission process to occur. The *activation energy* E_A can be estimated as the transformation between the barrier's extremity and the nucleus's energy when it is originally in the ground state. For instance, in a nucleus with $A \sim 240$, this energy is estimated to be 6 to 7 MeV.

It is important to note that the barrier must go through the quantum tunnel effect for the decay of the original nucleus by spontaneous fission. In order to understand the elements that govern the concern of channelling the completed barrier, it is remarkable to equate the Coulomb energies and surface of a spherical nucleus as anticipated by the well-known Bethe–Weizsacker semi-empirical mass formula that gives an admirable parameterization of the binding energies of nuclei in their ground state. This formula depends on the liquid-drop correlation; nevertheless, it also combines two quantum elements. First is *asymmetrical* energy that tends to turn constituents of the nucleus into an equal number of protons and neutrons. The further is *pairing* energy, which helps configure two indistinguishable fermions when forming the pairs. The Bethe and Weizsacker semi-empirical mass formula is,

$$B(A,Z) = a_v A - a_s A^{2/3} - a_c \frac{Z^2}{A^{1/3}} - a_a \frac{(N-Z)^2}{A} + \delta(A) \qquad \textbf{(2.1)}$$

The coefficients all together can be taken as 'a_i' (*i.e.* $a_i = a_v + a_s + a_c + a_a$) to govern the good estimation of experimental binding energies. The virtuous approximation is as follows,

a_v = 15.7 MeV (volume term)

a_s = 17.80 MeV (surface term)

a_c = 0.710 MeV (Coulomb term)

a_a = 23.69 MeV (asymmetry term)

And the quantum pairing term δ (A) can be given by,

$$\delta(A) = \begin{Bmatrix} 33.6\,A^{-3/4} & ,\text{if N and Z are Even} \\ -33.6\,A^{-3/4} & ,\text{if N and Z are Odd} \\ 0 & ,\text{if } A = N + Z \text{ is Odd} \end{Bmatrix} \quad \textbf{(2.2)}$$

The numerical values of all the above parameters essentially are calculated empirically (except coulomb repulsion term 'a_c'). On the other hand, the A and Z dependency of every single term reveal modest physical assets.

1.1.3.2. Photo−Fission

Photo fission is a progression in which a nucleus, subsequently absorbing a gamma-ray, experiences nuclear fission and disintegrates into two or more fragments. The fission percentage can be enhanced by engaging the nucleus in an excited state in order to decrease the fission barrier. This is ultimately completed by the photon (Gamma radiation of modest energies absorption in MeV). The photo fission can then convince fission in conventionally fissile elements, for instance, the actinides uranium, plutonium, and neptunium. Investigations have been accompanied by considerably sophisticated energy gamma-rays in the photo fission cross-section that change slightly inside and in the GeV range.

Photo transmutation or photodisintegration is analogous to photofission. However, a dissimilar physical practice to this is when an enormously high-energy gamma-ray act together with an atomic nucleus and causes it to pass in an excited state, which instantly falloffs by emitting subatomic particles. The following reaction can illustrate the fundamental photo fission process,

$$\gamma\,{}^{A}_{Z}X \rightarrow {}^{A}_{Z}X^{*} \rightarrow \text{Fission}$$

where AX* is any excited state of the original nucleus AX.

One of the major advantages of employing photo fission neutrons is that they are produced with around 200 times the yield of delayed neutrons; as a result, the probability of detection (PD) is higher and persists faster scan times.

Fig. (**2.6**) displays the overall photo-fission cross-section on ^{236}U as a role of photon energy.

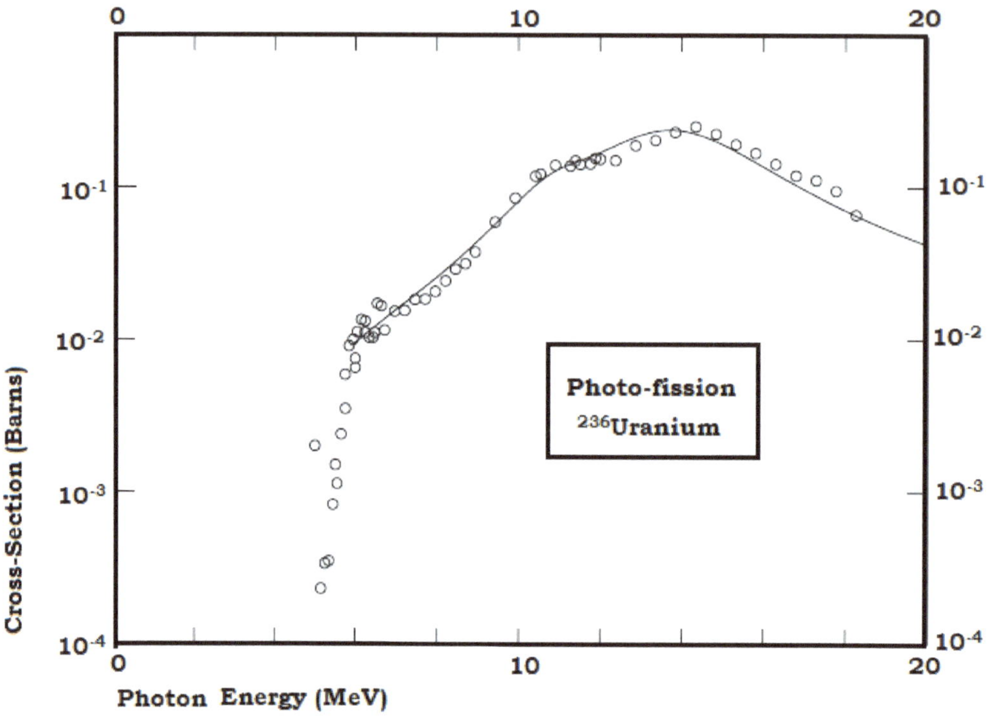

Fig. (2.6). The total photo-fission cross-section on ^{236}U.

The cross-section is insignificant for $E\gamma < 5$ MeV, quickly rises in the assortment 5 MeV $< E\gamma <$ 6 MeV and then gradually rises to its extreme value at $E\gamma \sim 14$ MeV. To a certain extent, we can randomly describe ΔES = 5.7 MeV, for instance, as the energy that essentially is supplemented to the ground state of ^{236}U to make the possibility for fission practically enormous [4]. This energy is predicted to be slightly less than the height of the barrier E_A.

$$\Delta E_S \sim E_A - 1 \text{ MeV}$$

This is because it is not essential to improve sufficient energy to remove the barrier but only adequate to generate the prompt tunnelling.

Fission energy at initial state (threshold) ΔE_S and neutron separation energy Sn for particular nuclei (*A, Z*). ΔE_S provides the actual inception for photofission. The actual initiation for neutron-encouraged fission of the nucleus ($A-1$) is $T_n = \Delta E_S - Sn$.

For the three odd-($A-1$) nuclei, threshold $T_n < 0$, fission can be induced by thermal neutrons. Table **2.2** grades the values of ΔE_S for particular nuclei in the expanse 230 < A < 240. They exist altogether in the order $\Delta E_S \sim 6$ MeV.

Table 2.2. Fission starting point energy ΔE_S and neutron departure energy Sn for nuclei (A, Z) and odd-(A − 1) nuclei.

Initial Fission Nucleus (A,Z)	Fission Threshold Energy 'ΔE_S'(MeV) (A,Z)	Separation Energy '*Sn*' (MeV) (A,Z)	Threshold 'T_n' (MeV) (A-1,Z)	Target Neutron (A-1,Z)
$^{234}_{92}U$	5.4	6.9	---	$^{233}_{92}U$
$^{236}_{92}U$	5.7	6.3	---	$^{235}_{92}U$
$^{240}_{94}Pu$	5.5	7.3	---	$^{239}_{94}Pu$
$^{233}_{90}Th$	6	5.1	1.3	$^{232}_{90}Th$
$^{235}_{92}U$	5.8	5.3	0.5	$^{234}_{92}U$
$^{239}_{92}U$	6.0	4.8	1.2	$^{238}_{92}U$

The last column of Table **2.2** provides the necessary information about the standards of the neutron-prompted fission inception for particular nuclei. Even-*N* nuclei have to start the kinematic energy of incoming neutrons. However, Odd -*N* target nuclei are fissionable by existing thermal neutrons ($T_n < 0$). Since the previous neutron of an odd-*N* fissionable nucleus is loosely bound as compared to the last neutron of an even-*N* fissionable nucleus, as mentioned in the pairing term [$\delta(A) = 34A^{-3/4}$] in the Bethe–Weizsacker semi-empirical mass formula (See section 1.6.3.2).

1.1.3.3. Fissile Materials and Fertile Materials

In the previous section, it has been perceived that for a nucleus with $A \approx 240$, the Coulomb barrier, which prevents unprompted fission, is between 5 and 6 MeV. If a neutron with null kinematic energy crosses a nucleus's threshold to form a composite nucleus, the latter will have excitation energy beyond its ground state identical to the neutron's binding energy in that state. For example, a zero-energy neutron entering a nucleus of ^{235}U generates a state of ^{236}U with an excitation energy of 6.3 MeV. This energy is right above the fission barrier, and the complex nucleus rapidly experiences fission, with deterioration yields comparable to those originating in the impulsive degeneration of ^{236}U. Alternatively, to encourage fission in ^{238}U, a neutron with a kinetic energy of at least 1.2 MeV is needed. The binding energy of the previous neutron in ^{239}U is only 4.8 MeV, and excitation energy of this extent is beneath the fission threshold of ^{239}U. The variances in the binding energies of the former neutron in even-A and odd-A nuclei are specified by the pairing term [1].

The nuclei which are utmost straightforwardly recycled as energy in fission receptacles are the three even-odd nuclei ^{233}U, ^{235}U and ^{239}Pu, which fission promptly by *thermal* neutron arrest. Surrounded by three fissile nuclides, only ^{235}U continues in remarkable expanses on Earth, which designates its antique prominence in the progress of nuclear information and expertise. Globally used uranium is (in the contemporary situation) a combination of isotopes encompassing 0.72% ^{235}U and 99.3% ^{238}U. Alternatively, ^{239}Pu and ^{233}U have α-decay lifetimes too little to be existing in global minerals. They are manufactured preciously by neutron arrest preparatory from the *fertile materials* ^{238}U and ^{232}Th:

$$n\,^{232}_{92}\mathrm{U} \rightarrow \,^{239}_{92}\mathrm{U}\gamma$$

$$^{239}_{92}\mathrm{U} \rightarrow \,^{232}_{92}\mathrm{Np}\ \overline{\nu e}\ e^{-} \qquad , t_{1/2} = 23.45\ \mathrm{m}$$

$$^{239}_{93}\mathrm{Np} \rightarrow \,^{239}_{94}\mathrm{Pu}\ \overline{\nu e} \qquad , t_{1/2} = 2.3565\ \mathrm{days}$$

And

$$n\,^{232}_{90}\mathrm{Th} \rightarrow \,^{233}_{90}\mathrm{Th}\gamma$$

$$^{233}_{90}\mathrm{Th} \rightarrow \,^{233}_{91}\mathrm{Pa}\ \overline{\nu e}\ e^{-} \qquad , t_{1/2} = 22.3\ \mathrm{m}$$

$$^{233}_{91}\mathrm{Pa} \rightarrow \,^{233}_{92}\mathrm{U}\ \overline{\nu e} \qquad , t_{1/2} = 26.967\ \mathrm{days}$$

1.1.4. Moderation and Thermalization of Neutron

One of the scientifically most significant interactions of neutrons with matter is their loss of energy (“slowing down”) by a chain of elastic collisions. These can be preserved by the approaches of classical mechanics, supposing the interacting particles as flawless elastic spheres.

1.1.4.1. Moderation of Neutron

Neutron moderation is the procedure by which one can eliminate approximately 99.99999% of the kinematic driving energy of the preliminary or ultimate neutrons, carrying them to convenient energy utilization.

It is essential to secure a short-pulsed source like a Slow Neutron Source (SNS) without terminating the sharp emission time using the time-of-flight technique. The slowing-down theory pronounces the loss of energy by repetitive elastic collisions with effectually free nuclei [2].

1.1.4.2. Thermalization and Slowing Down of Neutron

Fission neutrons are created at an ordinary energy level of 2 MeV and instantaneously trigger to downcast, resulting in consequential scattering reactions with a range of aimed (target) nuclei. Afterward, the total extent of collisions with nuclei, the velocity of a neutron is decreased to such an amount that it has almost the equivalent typical kinematic energy as the atoms (or molecules) of the medium in which the neutron experiences mutable smattering called elastic scattering. This energy, which is merely a minor constituent of an electron volt at average temperatures (0.025 eV at 200 ^{0}C), is commonly mentioned as thermal energy determined by the temperature. The energies of neutrons that have been depressed to the amounts (< 1 eV) in this region are nominated as thermal neutrons. This kind of phenomenon of elastic scattering in which the energy of a neutron decreases to the thermal region is considered slowing down, moderation or thermalization. The matter or a substance used for the persistence of thermalizing neutrons is permitted as a moderator.

An ideal moderator decreases the velocity (deacceleration) of neutrons in an insignificant amount of collisions but does not captivate them to an excessive amount. The ultimate controlling material (moderator) should have the resulting nuclear properties as itemized beneath,

i. enormous scattering cross-section
ii. slight absorption cross-section
iii. great energy loss for each collision

The deceleration of neutrons in an insignificant amount of collisions is imperative to decreasing the number of neutrons escaping from the interior of the nucleus and correspondingly decreasing the amount of absorption in non-fuel constituents.

To slow down neutrons, it is necessary to bounce them off, which is possible by turning the kinetic energy of neutrons into recoil energy. To do so, the effective energy loss parameters should be taken into consideration as,

i. Smallest portion leftward when a collision is over: Energy-loss ratio 'α.'
ii. Slight loss during each collision: Reduction in average logarithmic energy 'ξ'
iii. Neutrons escape from the moderator as they relax and eventually deject emerging neutron beams
iv. The rate of collisions is impartially momentous in slowing down as the amount of energy vanished during each collision
v. Lighter nuclei need a reduced number of collisions to slow the neutron
vi. Extraordinary macroscopic scattering cross-section for faster collision rate originates from both number density and microscopic cross-section

An appropriate amount of energy annihilation during each collision is the reduction in logarithmic energy. The usual reduction in logarithmic energy is equivalent to the usual decrease in neutron energy for each collision in the logarithm, characterized by the symbol 'ξ'

$$\xi = \log E_I - \log E_F = \log\frac{E_I}{E_F} \quad \textbf{(2.3)}$$

Where, E_i= the average initial energy of neutron and

E_f = the average final energy of the neutron

As the element of energy recollected by a neutron in a particular elastic collision is persistent for a specified substance, the neutron's energy decrease for each collision 'ξ' will be constant. Because it is endless for all kinds of physical entities and is not influenced by the preliminary neutron energy, ξ is an appropriate magnitude for determining the moderating capacity of a quantifiable matter.

Therefore the complete collision (N) essential for a neutron to utilize a distinct extent of energy might be resolute by dividing 'ξ' into the transformation of the natural logarithms of the energy array in interrogation. The amount of collisions (N) that are transferrable from a higher energy (E_H) to a lower energy (E_L) can be estimated as follow,

$$N = \frac{\log E_H - \log E_L}{\xi} = \frac{\log\left(\frac{E_H}{E_L}\right)}{\xi} \quad \textbf{(2.4)}$$

Occasionally it becomes appropriate to establish the statistics from known proceedings, to start with a typical minuscule energy loss for each collision using an average logarithmic fraction [3]. Then the ending energy for a particular figure of collisions may be deliberated using the subsequent formulation 2.5,

$$E_{F(NEUTRON)} = E_{I(NEUTRON)}\{1 - X\}^N \quad \textbf{(2.5)}$$

Where, $E_{I\ (NEUTRON)}$ = the initial energy of neutron,

$E_{F\ (NEUTRON)}$ = the final energy for a specified number of collisions, and

'X'= the typical fractional energy loss per collision,

1.1.4.3. Practical Slowing-Down of Neutron

The downfall in energy during each collision decreases with the mass of the nucleus, and the percentage of collisions increases with advanced scattering cross-section. Various parameters that control the slowdown process of the neutron are tabulated in Table **2.3**.

Table 2.3. Various parameters responsible for slowing down the neutron.

Moderator	Energy-loss Ratio 'α'	Reduction in Average Logarithmic Energy 'ξ'	Number of Collisions to Thermalize (up to 1eV)	Time Required to 1eV (μs)
Water (H_2O)	-	0.93	16	1.5
Heavy Water(D_2O)	-	0.51	28	9.7
Beryllium (Be)	0.64	0.21	69	8.5
Helium(He)	-	0.42	42	-
Carbon (C)	0.72	0.16	91	25
Boron(B)	-	0.17	105	-
Iron (Fe)	0.93	0.04	414	43
Lead (Pb)	0.98	0.01	1450	390

From Table **2.3**, it is clear that the faster the slowdown rate of neutrons, the lesser the initial contracted time of scattering will be to degrade the neutron for which the accelerator fellows work effortlessly. Water, heavy water and beryllium are considered good moderators [4].

1.1.4.4. Slowing-Down Units

Removing approximately 99.9999% of neutron energy in the slowdown process is a logarithmic function of loss of energy and expressed as Neutron Lethargy (u) in general,

$$u = -\log\frac{E_{max}}{E} \qquad \textbf{(2.6)}$$

From above equation 2.6, it can be concluded that,

i. Dropping 99.9999% of energy (E) is only acquiring 13.8 lethargy units
ii. Dropping 99.999999% of energy (E) means acquiring 18.4 lethargy units

In general, only crystallography marks enhanced logic in reciprocal space; neutron physics is informal in lethargy space. In addition to lethargy, sometimes the other communal units like energy (E), wavelength (λ), frequency (ν), velocity (v), and wave vector (k) can also be used.

For example: 1 Å ≈ 0.0818 eV ≈ 3956 m/s ≈ 6.28 1/Å

In the above framework, Temperature (T) is not the same parameter because it relates to a scattering of neutrons and not an individual neutron. Slowing-down of neutrons practically points to 1/E of the spectrum (continuum) indicated in Fig. (**2.7**).

For example: φ (E) ≈ 1/E

Or φ (u) ≈ Eφ (E) = ½λφ (λ) ≈ constant.

Where φ = flux proportion of neutrons of entire dynamic energies transmitting from side to side of sphere of elemental cross-sectional area.

1.1.4.5. Neutron Transport in Matter

A comprehensive neutron transport analysis is outlying away from the possibility of the present transcript. It is together essential in nuclear expertise and very complex to resolve. The transport calculations are an integral as well as differential equivalence whose mathematical conduct is in the aforementioned originality that has been progressively established for years in almost entire nuclear investigation platforms. Its convolution originates in fragment commencing the measurement that it indulgences the performance of neutrons mutually as a role of energy as they can drop energy in collisions and space as they distribute. The investigation of neutron transport can also be resolved by the Boltzmann statistics.

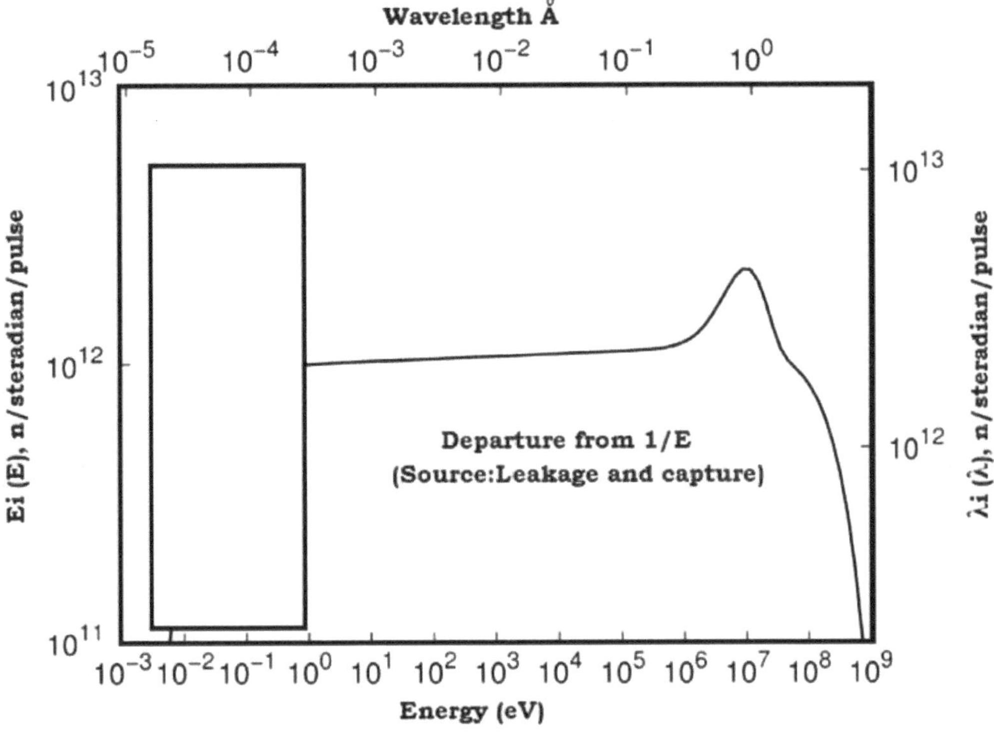

Fig. (2.7). Source and Slowing Down Spectrum of Neutron.

In the particular state of a neutron, the Boltzmann statistics can resolve the problem using very simple calculations. But then again, it is important to determine why and by what means the perception of a "critical mass" arises. The difficulty is relatively unassuming if we create the thoroughly irrational expectations that the neutrons all have identical time-independent energy and that the medium is also identical, although constrained in magnitude [5, 6]. Henceforth the Boltzmann statistics gives directions for the actions of neutrons in the matter, and certainly, it is based on subsequent conventions:

i. The medium is stationary (avoiding insignificant thermal signals), and it is spherical and homogeneous;

ii. Neutron–neutron scattering is insignificant as the density of neutrons is considerably reduced as compared to the density of the stationary medium;

iii. Neutron disintegration is insignificant because the neutron lifetime is enormous as matched to the usual time alterations between two exchanges.

By considering the above conventions, the neutrons can be designated by their density in phase space,

$$f(r,p,t) = \frac{dN}{d^3pd^3r} \quad \textbf{(2.7)}$$

where, d^3pd^3r = phase space element

dN = the number of neutrons in phase space.

The density of space of particular neutrons and the flow of charges recounting the spatial flow of neutrons are the integral function momentum (p) as,

$$\int f(r,p,t)\, d^3p = n(r,t) \quad \textbf{(2.8)}$$

$$\int vf\,(r,p,t) d^3p = J(r,t) \quad \textbf{(2.9)}$$

When collisions are completely absent, the momentum of neutrons becomes time-independent, and the movement of elements in phase space is created by considering the particle's motion at velocities v = p/m. The density f can be estimated by the equation,

$$C(f) = \frac{\partial f}{\partial t} + v.\Delta f \quad \textbf{(2.10)}$$

where C (f) is the term that originated from collision progressions but in the absence of collision C (f) = 0, and above equation (2.10) will become,

$$\frac{\partial f}{\partial t} + v.\Delta f = 0 \quad \textbf{(2.11)}$$

The above equation (2.11) is entitled the *Liouville equation.*

The rates of elastic and absorption scattering (λ_{ELA} *and* λ_{ABS}) are products of the fundamental cross-sections, the concentration of scattering centres $n_{239,}$ and the average velocity ‘v,’

$$\lambda_{ELA} = v n_{239} \sigma_{ELA}$$

$$\lambda_{ABS} = v n_{239} \sigma_{ABS}$$

Here the absorption arises from fission reaction and (n,γ) interactions. And it is well known that the origin of both the interaction is an elastic collision,

The collision term can be elaborated as follow,

$$C\{f(p)\} = n_{239} \int d^3\, p' v(p')\, f(r, p', t) \frac{d\sigma}{d^3 p'} (p' \rightarrow p) - [\lambda_{ELA} + \lambda_{ABS}]\, f(r, p, t) + S(r, p) \quad \textbf{(2.12)}$$

The principle term on the left-hand side of the above equation interprets the fundamental interactions in terms of elastic collision, whereas the initial term on the right-hand side represents the neutrons approaching from the components of phase space d^3rd^3p', which cross the threshold element of phase space d^3rd^3p using elastic scattering. The intermediate-term on the right-hand side signifies the neutrons that leave the component d^3rd^3p either by elastic scattering or by absorption, and the last term, S(r, p), is a basic term denoting the generation of neutrons in the fission process [7].

1.2. Nuclear Fusion

The progression in which two or more light nuclei are pooled together into a solitary nucleus with the discharge of marvellous energy is called nuclear fusion.

Resembling a fission reaction, the summation of masses before the fusion (*i.e.*, of light nuclei) is more than the summation of masses after the fusion (*i.e.*, of a greater nucleus), and this alteration gives an impression just like fusion energy. The most characteristic fusion reaction is the fusion of two deuterium nuclei into helium, given below.

A deuterium nucleus and a tritium nucleus combine to form a helium-4 (^{4}He) nucleus [8] and a neutron. Agreeing to Einstein's Mass-Energy co-relation ($E=mc^2$), the difference in the masses is transformed into the kinetic energy of the emerging particle.

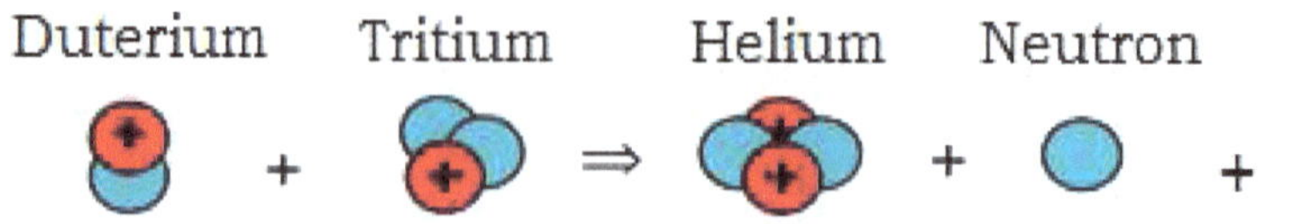

Duterium		Tritium		Helium		Neutron		Energy in MeV
	+		⇒		+		+	
$_1H^2$	+	$_1H^3$	→	$_2He^4$	+	$_0n^1$		

Masses

$_1H^2$	2.014102 u	$_2He^4$	4.002603 u
$_1H^3$	3.016050 u	$_0n^1$	1.008665 u
	5.030152 u		5.011268 u

For the fusion response to materialize, the light nuclei are conveyed nearer to each other (with a distance of 10^{-14} m). This is probable only at a very high temperature to counter the repulsive force stuck between nuclei. Nuclear fusion needs tremendously high temperatures. This is because the small nuclei necessitate sufficient kinetic energy to overwhelm their electrostatic repulsion. The energy we obtain from the sun is from nuclear fusion. The sun is made up mostly of hydrogen and helium. In the sun's interior, the temperature is millions of degrees Celsius, and there is the continuous fusion of small nuclei into larger nuclei [9, 10].

The discharge of energy using the fusion of light elements is owed to the interaction of two different forces: the nuclear force, which groups composed of protons and neutrons, and the Coulomb force, which origins protons to keep away from each other. Protons are positively charged and kept away from each other by the Coulomb force; nevertheless, they can have altercations with each other, representing the presence of other entities, and short-range force signifies nuclear attraction [11]. Nuclei smaller than iron and nickel, called light nuclei, are adequately small and proton-deprived, which consent the nuclear force to controlled revulsion. This is because the nucleus is appropriately small that all nucleons sense the short-range attractive force as forcefully as they experience the infinite-range Coulomb repulsion. Constructing up nuclei from lighter nuclei by fusion liberates the superfluous energy from the total attraction of particles. However, no energy is obtainable for bigger nuclei, as the nuclear force is short-range and terminates to act across extended nuclear length processes. Consequently, energy is not on the loose with the fusion of such nuclei; in its place, energy is indispensable as involvement for such improvements.

Fusion controls stars and yields all components in a stellar nucleosynthesis process. The Sun is the main star and yields its energy by nuclear fusion by adapting hydrogen nuclei into helium. In its core, the Sun fuses 620 million metric tons of hydrogen and generates 616 million metric tons of helium at the individual moment of fusion. The fusion of lighter components in stars issues energy and the mass that continuously rises it. For instance, in the assimilation of two hydrogen nuclei to produce helium, 0.645% of the mass is transported away in the form of the kinetic energy of an alpha particle or new forms of energy, such as electromagnetic radiation [12]. Fig. (**2.8**) indicates the representation of basic nuclear reaction and emission of energy.

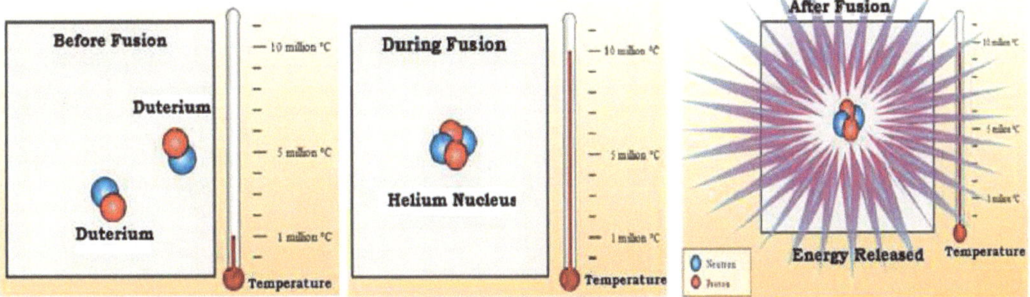

Fig. (2.8). Basic Nuclear Fusion Reaction and emission of energy.

1.2.1. Elementary Reaction for Energy Discharge in the Sun by Fusion

The process which occurs at the core of the sun is also the Fusion reaction. The light emitted by the sun and the warmth we feel as a result of it is also the result of a fusion reaction. When hydrogen nuclei collide, they fuse into heavier helium atoms, and a significant quantity of energy is released in this process shown in Fig. (**2.9a**) below.

1.2.2. Elementary Reaction for Energy Discharge in the Sun by Fusion

In the sun, energy production is a multi-phase development. There are 4 entire phases elaborate on the energy discharged from the sun's core.

Phase 1:

$$^{11}\mathrm{H}\ (\text{proton}) + {}^{11}\mathrm{H}\ (\text{proton}) \rightarrow {}^{21}\mathrm{H}\ (\text{deuteron}) + e^{+}(\text{positron}) + \nu\ (\text{neutrino}) + 0.42\ \text{MeV}$$

Phase 2:

$$e^{+} \text{ (positron)} + e^{-} \text{ (electron)} \rightarrow \gamma \text{ (Gamma rays)} + 1.02 \text{ MeV}$$

Phase 3:

$$^{21}\text{H (deuteron)} + {}^{11}\text{H (proton)} \rightarrow {}^{32}\text{He (helium)} + \gamma \text{ (Gamma rays)} + 5.49 \text{ MeV}$$

Phase 4:

$$^{32}\text{H} + {}^{32}\text{H} \rightarrow {}^{42}\text{He} + {}^{11}\text{H} + {}^{11}\text{H} + 12.86 \text{ MeV}$$

Phases 1, 2 and 3 come about twice in the sun, and phase 4 takes place merely once.

As soon as all the above 4 reactions are pooled together, four hydrogen atoms associate to produce a ^{42}He atom with a discharge of 26.7 MeV of energy.

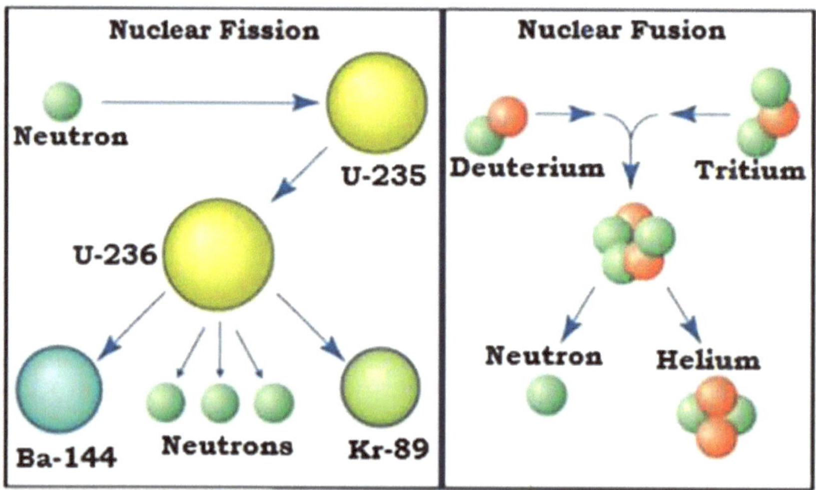

Fig. (2.9). (a): Basic Mechanism between Fission and Fusion process.

Concluding reaction is:-

$$^{41}\text{H} + 2e^{-} \rightarrow {}^{42}\text{He} + 6\gamma + 2\nu + 26.7 \text{ MeV}$$

It is correspondingly recognized as a proton-proton cycle because this improvement flinches with protons [13].

REFERENCES

[1] Walter, Marni Blake (2015-09-01). Western Pennsylvania History Magazine. 98 (3): 36–49. Retrieved 2019-12-03.

[2] Haxby, R.O.; Shoupp, W.E.; Stephens, W.E.; Wells, W.H. (1941-01-01). "Photo-Fission of Uranium and Thorium". *Physical Review*. 59 (1): 57–62. Bibcode:1941PhRv...59...57H. [http://dx.doi.org/10.1103/PhysRev.59.57]

[3] J.A. Silano, and H.J. Karwowski, "Near-barrier photofission in Th 232 and U 238", *Phys. Rev. C,* vol. 98, no. 5, p. 054609, 2018. [http://dx.doi.org/10.1103/PhysRevC.98.054609]

[4] D. Doré, J-C. David, M-L. Giacri, J-M. Laborie, X. Ledoux, M. Petit, D. Ridikas, and A.V. Lauwe, "Delayed neutron yields and spectra from photofission of actinides with bremsstrahlung photons below 20 MeV", *J. Phys. Conf. Ser.,* vol. 41, pp. 241-247, 2006. [http://dx.doi.org/10.1088/1742-6596/41/1/025]

[5] P.J. Mohr, and B.N. Taylor, "CODATA recommended values of the fundamental physical constants: 1998", *Rev. Mod. Phys.,* vol. 72, no. 2, pp. 351-495, 2000. [http://dx.doi.org/10.1103/RevModPhys.72.351]

[6] E.G. Kessler Jr, M.S. Dewey, R.D. Deslattes, A. Henins, H.G. Börner, M. Jentschel, C. Doll, and H. Lehmann, "The deuteron binding energy and the neutron mass", *Phys. Lett. A,* vol. 255, no. 4-6, pp. 221-229, 1999. [http://dx.doi.org/10.1016/S0375-9601(99)00078-X]

[7] B. Franzke, "The heavy ion storage and cooler ring project ESR at GSI", *Nucl. Instrum. Methods Phys. Res. B,* vol. 24-25, no. 25, pp. 18-25, 1987. [http://dx.doi.org/10.1016/0168-583X(87)90583-0]

[8] G. Savard, and G. Werth, "Precision Nuclear Measurements with Ion Traps", *Annu. Rev. Nucl. Part. Sci.,* vol. 50, no. 1, pp. 119-152, 2000. [http://dx.doi.org/10.1146/annurev.nucl.50.1.119]

[9] P. De Bievre, S. Valkiers, R. Gonfiantini, P.D.P. Taylor, H. Bettin, F. Spieweck, A. Peuto, S. Pettorruso, M. Mosca, K. Fujii, M. Tanaka, Y. Nezu, A.J. Leistner, and W.J. Giardini, "The molar volume of silicon [Avogadro constant]", *IEEE Trans. Instrum. Meas.,* vol. 46, no. 2, pp. 592-595, 1997. [http://dx.doi.org/10.1109/19.571927]

[10] J.K. Dickens, F.G. Perey, and R.J. Silva, "Level Structure of Ni 64", *Phys. Rev.,* vol. 132, no. 3, pp. 1190-1192, 1963. [http://dx.doi.org/10.1103/PhysRev.132.1190]

[11] Koga Toshikatsu, and Koshida Yoshinori, " Roles of nuclear attraction and electron repulsion energies in the relative stability of atomic LS terms", *J. Chem. Phys.,* vol. 111, no. 54, 1999. [http://dx.doi.org/10.1063/1.479253]

[12] H.A. Bethe, "The Hydrogen Bomb", *Bull. At. Sci.,* vol. 6, no. 4, pp. 99-104, 1950. [http://dx.doi.org/10.1080/00963402.1950.11461231]

[13] J. Sharpey-Schafer, "Renaissance of nuclear spectroscopy", *Phys. World,* vol. 3, no. 9, pp. 31-35, 1990. [http://dx.doi.org/10.1088/2058-7058/3/9/24]

CHAPTER 3

Nuclear Structure and Properties of Nuclei

Abstract: An atom's nucleus comprises neutrons and protons, which in turn are the appearance of more fundamental particles, called quarks, that are seized in a relationship by the strong nuclear force in certain stable arrangements of hadrons, called baryons. The strong nuclear force encompasses far enough from each baryon to drag the neutrons and protons together beside the repulsive electrical force between the positively charged protons. The present chapter deals with investigating and correlating key features of the nuclear structure and its properties as understanding the structure of the atomic nucleus is one of the central challenges in nuclear physics. The line of understanding in this chapter is subjected to the structure of nuclei, atomic models, Rutherford model of the atom, nuclear composition, nuclear properties, determination of mass and determination of the charge.

Keywords: Charge, Mass, Nuclear Properties, Nuclear Structure, Nuclear Structure.

1. INTRODUCTION

When discussing the atomic nucleus, it is very important to know about the history of the atom. An 'Atom' is the smallest particle of a substance that can exist by itself. Each atom consists of a nucleus with a positive charge and a set of electrons that move around the nucleus. The atom in its normal state is always found to be electrically neutral so that the number of protons and electrons must be exactly equal.

The nucleus was revealed in 1911 due to Ernest Rutherford's determinations to test Thomson's "plum pudding model" of the atom [1]. J.J. Thomson had previously revealed the electron and deliberated that atoms are electrically neutral; he hypothesized that there must also be a positive charge. In his desirable pudding model, Thomson recommended that an atom comprised of negative electrons is arbitrarily dispersed inside a sphere of positive charge. Ernest Rutherford later developed experimentation with his research mate Hans Geiger and with the help of Ernest Marsden, which elaborated the deflection of alpha particles (helium nuclei) focused on a thin sheet of metal foil. He well-structured that if J.J Thomson's model were accurate, the positively charged alpha particles would certainly pass through the foil with very slight deviance in their paths, as the foil

Ritesh Kohale, Sanjay J. Dhoble & Vibha Chopra

should act as electrically impartial if the negative and positive charges are so confidentially mixed as to make it perform neutral. To his disclosure, many particles were deflected at great angles. Because the mass of an alpha particle is about 8000 times that of an electron, it became obvious that a very strong force must exist if it could deflect the enormous and fast-moving alpha particles. He recognized that the desirable pudding model could not be precise and that the deflections of the alpha particles could only be clarified if the positive and negative charges were disconnected from each other and that the atom's mass was a concentrated point positive charge. This warranted the idea of a nuclear atom with a dense center of positive charge and mass.

The term nucleus is from the Latin word nucleus, a tiny part of nux ('nut'), which signifies the grain (*i.e.*, the 'small nut') within a damp type of fruit (like a peach). In 1844, Michael Faraday recycled the term to denote the "central point of an atom." The modern atomic significance was projected by Ernest Rutherford in 1912 [2]. Nevertheless, the assumption of the term "nucleus" in atomic theory was not instantaneous. In 1916, for example, Gilbert N. Lewis stated, in his well-known piece of writing "The Atom and the Molecule," illuminating the concept that "the atom is composed of the grain and an outer atom or shell" [3].

1.1. Atomic Models

Different models were proposed to explain an atom. In 1808, John Dalton published his theory suggesting that the atom is the smallest unit of matter. It can neither be created nor be destroyed. Later in 1897, the 'Plum Pudding model' or 'watermelon model' was put forward by Sir J.J. Thomson. It was suggested that the atom consisted of a positive electric fluid distributed over a whole body of atoms, and negative electrons are embedded in it like seeds in watermelon (Fig **3.1**).

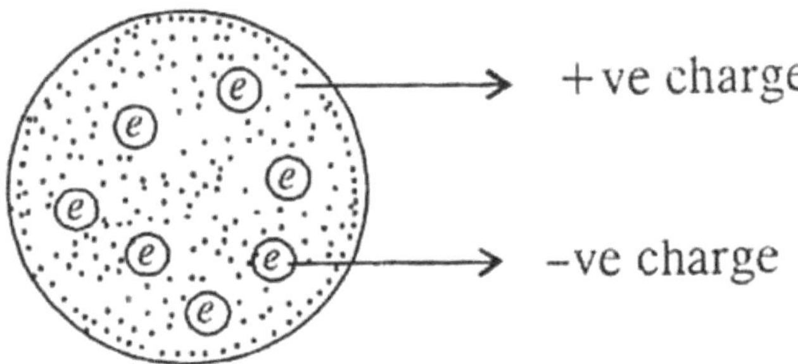

Fig. (3.1). Schematics of an atom as the smallest unit of matter.

1.1.1. Rutherford Model of the Atom

J.J. Thomson's model of the atom was static. Later this static model was transformed into a dynamic 'Planetary Model' in 1911 by Rutherford as the static model of the atom failed to explain the α-scattering experiment performed by Rutherford [4]. The schematic diagram of Rutherford's experimental setup is shown in Fig. (**3.2**).

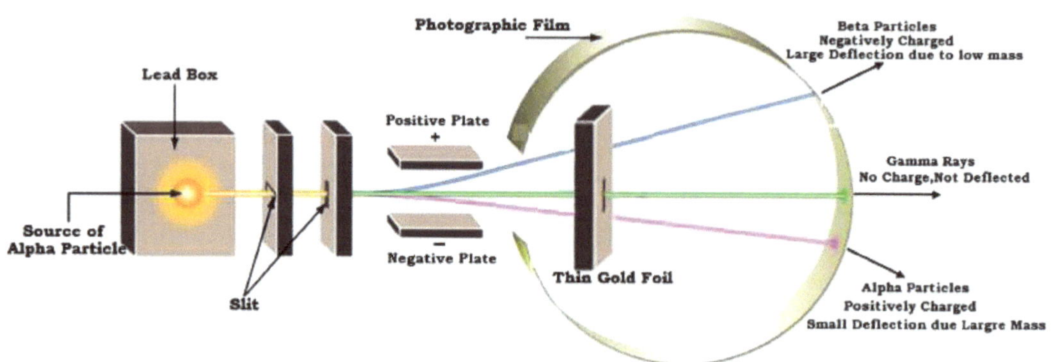

Fig. (3.2). Rutherford's gold foil experiment. (Source Credit: Britannica).

A beam of ∝-particles from the Radium source was made to strike a thin gold foil. A fluorescent screen detected the scattered s-particles. Rutherford observed that many ∝-particles went straight through the foil or were deflected at very small angles. However, few of the ∝-particles were deflected through very large angles, and few were scattered back. J.J. Thomson's model of the atom could not explain the deflections. So, to explain the deflections, Rutherford assumed that the positive massive part of the atom was concentrated in a very small volume at the centre of the atom called the nucleus. This nucleus is surrounded by an electron cloud, making the entire atom electrically neutral.

An ∝-particle is scattered at the larger angle because an ∝-particle approaching the centre of the atom experiences an increasingly large Coulomb repulsion. So, the nucleus was assumed to have a negative charge. Most ∝-particles do not go sufficiently close to the nucleus, so they pass through the foil with practically no deviation, so the atom was assumed to have a lot of empty space. Further, it was assumed that there were enough electrons outside the positive nucleus to provide a balance of positive and negative charges.

Despite all this, this model also failed. The atoms did not stabilize through electrostatic forces alone since the extranuclear electrons were stationary. This created a problem even if the electrons were assumed to revolve around the nucleus, as the planets revolved around the sun, and the centrifugal force prevented the electrons from falling into the nucleus. According to classical electromagnetic theory, an accelerated electron should radiate energy. So, the electron will fall into the nucleus *via* a spiral path, coming closer and closer to the nucleus as it loses energy [5, 6].

Niels Bohr overcame this situation in 1912 by introducing Bohr's model of the atom and the concept of non-radiating quantized electron orbits. So, after this, some other models were proposed, like the Sommerfeld vector atom model in 1915, to explain the atomic structure of the atom.

1.2. Nuclear Composition

Rutherford's scattering experiment established the validity of the concept of the nucleus of an atom. It was also clear that the atom as a whole is electrically neutral, with equal positive and negative charges. However, the distribution of these charges in an atom was not clear.

Further, the phenomenon of natural radioactivity suggests that the compact nuclear structure is capable of emitting α, β and γ rays. Gamma (γ) rays are not the constituents of the nucleus. Rather, they are emitted when a nucleus returns from an excited state to a normal state. Moreover, it was established that $\propto$-particles, protons, neutrons and electrons form the nucleus's constituents. So, based on the experimental facts, certain theories were proposed to explain the nuclear composition. Two important theories are (p-e) theory and (p-n) theory, which are as follows [7].

1.2.1. Proton-electron (P-e) Theory

According to proton-electron theory, the nucleus was supposed to consist of protons and electrons. The reason behind such a concept was an experimental fact: electrons are emitted by radioactive nuclei in the β-decay process. Also, it was suggested from mass spectrum data that the masses of different nuclei were whole number multiples of the mass of hydrogen, and the hydrogen nucleus was given the name proton. But actually, the whole number of atomic weights was first an approximation; most of the atomic weights were fractional due to the presence of two or more isotopes.

So, to justify the nucleus mass with an atomic weight close to integer A, it was assumed that nuclear contains (A) protons. It means the charge on the nucleus should be (A) and not equal to the atomic number (Z). So, to overcome the difficulty, it was assumed that since the atom is electrically neutral with the protons, the nucleus contains (A-Z) electrons to make the charge of the nucleus equal to +Z as required, and further, the nucleus was assumed to be surrounded by Z extra nuclear electrons, *e.g.*, carbon was considered to have 12*p* and 12*e* [8]. Six electrons were considered inside the nucleus and 6 outside the nucleus, so the atom as a whole is neutral.

(i) Failures of (*p-e*) Theory

Although the (*p-e*) theory had some satisfactory aspects, there were still strong reasons which ruled out the possibility of electrons inside the nucleus. A few of such reasons are as follows.

(a) De Broglie Wavelength It was assumed that if the electrons exist inside the nucleus, their De Broglie wavelength should not exceed the size of the nucleus. For electron,

$$\text{Mass of electron } m_e = 9 \text{ x } 10^{-31} \text{ kg}$$

De Broglie wavelength for the electron (d) = $\frac{h}{m_e c}$

$$= \frac{6.6\,3 \text{ x } 10^{-3} \text{J.S}}{9\text{x}10^{-31} \text{ x } 3 \text{ x } 10^{8}\frac{\text{m}}{\text{s}}}$$

$$= \frac{6.6\,3 \text{ x } 10^{-34}}{27 \text{ x } 10^{-23}} \text{ m}$$

$$= \frac{6.6\,3}{27} \text{ x } 10^{-11}\text{m}$$

$$= 0.24 \text{ x } 10^{-11}\text{m}$$

The De Broglie wavelength for electrons comes out to be an order of 10-11m, whereas the nuclear radius is 10-^{14}m. Since De Broglie's wavelength comes out to be 1000 times greater than the size of the nucleus, so the assumption was wrong; hence electrons cannot exist inside the nucleus.

(b) Wave Mechanical Consideration

According to Heisenberg's uncertainty principle,

$$\Delta x.\ \Delta p \sim h.$$

if the principle is applied to an electron in the nucleus, then the uncertainty in the position (x) is Δx = diameter of the nucleus

$$\Delta x = 2 \text{ x } 10^{-14}\text{m}$$

So uncertainty in momentum

$$= \frac{6.6\ 3 \text{ x } 10^{-14}}{2 \text{ x } 3.14 \text{ x } 2 \text{ x } 10^{-14}}$$

$$= \frac{1.05}{2} \text{ x } \frac{10^{-34}}{10-14}$$

$$= 0.5 \text{ x } 10^{-20} \text{ Jsm}^{-1}$$

$$\Delta \text{p} \approx 10^{-20} \text{ Jsm}^{-1}$$

The total energy of an electron

$$\text{E}^2 = \text{p}^2\text{c}^2 + \text{m}_0^2\ \text{c}^4$$

$$\text{E} = \sqrt{10^{-40} \text{ x } 9 \text{ x } 10^{16} + (9 \text{ x } 10^{-31} \text{ x}}\ (3 \text{ x } 10^8)^2)^2$$

$$\text{E} = \sqrt{9(10)^{-24} + 6.6 \text{ x } 10^{-30}\ \text{J}}$$

$$\text{E} = \frac{3 \text{ x } 10^{-12}}{1.6 \text{ x } 10^{-19}}$$

$$\approx 20 \text{ MeV}$$

So the free-electron confined within the nucleus should have a kinetic energy of the order of 20 MeV. But experimentally, the electrons emitted by radio nuclei can never have K.E greater than about 4 MeV. Therefore electrons cannot exist inside the nucleus.

(c) Nuclear Spin

Each electron and proton have a spin of h2; hence the nucleus must have an integral spin if it contains total even number and a half-integral spin if it contains a total odd number of electrons and protons inside the nucleus. But experimentally, it has been found that spin depends on the mass number A of the nucleus zero or an integer for even A and is half-integral for odd A. And the number of particles inside the nucleus will be 2A-Z. *E.g.*, the Beryllium nucleus, $_4Be^{9,}$ contains 9 protons and 5 electrons, the total number of particles inside the nucleus is 14, so its total spin should be integral. But experimentally, it was found to have a half-integral spin of magnitude 3h2. So, it ruled out the existence of electrons inside the nucleus [9, 10].

(d) Nuclear Magnetic Moment

The magnetic moment of an electron is

$$\mu = \frac{eh}{2me} Am^2$$

$$\mu = \frac{eh}{2\,x\,9.1\,x\,10^{-31}}$$

$$\mu = 1\,x\,10^{30}\left(\frac{eh}{2}\right) Am^2$$

Magneto moment of photon

$$\mu = \frac{eh}{2m_p} = \frac{eh}{2\,x\,1.67\,x\,10^{-27}}$$

$$\mu = \left(\frac{eh}{2}\right)\left(\frac{1}{2}x\,10^{27}\right) Am^2$$

$$\mu = 2000$$

So if electrons exist inside the nucleus, then the electronic magnetic moment must dominate the nuclear magnetic moment. Experiments show that nuclear magnetic moment is quite small, meaning electrons have not contributed to it. Consequently, electrons cannot exist inside the nucleus.

(e) Finite-size of the Electron

If the electrons are assumed to be a sphere of finite dimensions, therefore in the case of heavy elements, the accommodation of a large number of electrons in the nucleus will be difficult as the dimensions of such a nucleus were found to be bigger than the experimentally measured values. So, it proved that electrons could not exist inside the nucleus.

(f) β^+ Decay

Proton-electron theory could not explain β^+ decay, *i.e.*, emission of positrons, because if both electrons and positions exist inside the nucleus, their annihilation occurs.

(g) Statistics

Proton-electron theory could not justify the statistics of a few elements, *e.g.*, in the case of ${}_7^{14}N$, it consists of 2A-Z = 21 fermions, so it should obey Fermi-Dirac statistics, but the band spectra and Raman spectra of ${}_7^{14}N$ shows the Bose-Einstein statistics, indicating that electrons don't exist inside the nucleus.

1.2.2. Proton-neutron (p-n) Hypothesis

After the experimental discovery of the neutron in 1932, Heisenberg proposed a (*p-n*) hypothesis that nuclei consist of protons and electrons. Also, according to this hypothesis, the total number of particles inside the nucleus (protons + neutrons) defines the mass number A. The number of protons gives the atomic number Z, and the number of neutrons is N = A-Z. This atom is composed of protons, neutrons and electrons. (Shown in Fig. (**3.3**)).

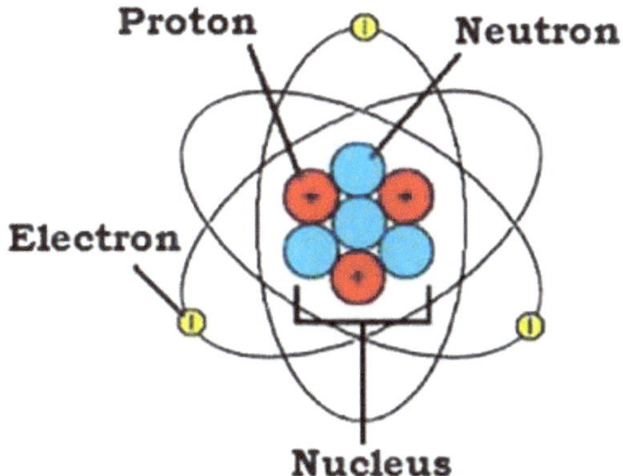

Fig. (3.3). Atomic model composed of protons, neutrons and electrons.

The protons and neutrons inside a nucleus are given a common name, nucleon, and the electrons form the extranuclear structure of an atom. Further (*p-n*) hypothesis overcomes the problems of the (*p-e*) theory of nuclear composition that is discussed below:

(a) Wave Mechanical Consideration: If the nucleus contains neutrons and Δx = diameter of the nucleus

$$\Delta x \approx 2 \text{ x } 10^{-14}\text{m}$$

Then uncertainty in momentum is

$$\Delta p = \frac{h}{\Delta x} = \frac{\text{h}}{2\mu \text{ x } 2 \text{ x } 10^{-14}}$$

$$\Delta p \approx 10^{-20} Jsm^{-1}$$

Total energy of neutron (or proton)

$$\text{E} = \sqrt{\text{p}^2\text{c}^2 + \text{m}_0^2\text{c}^4}$$

$$= \sqrt{10^{-40} \text{ x } 9 \text{ x } 10^{16} + [1.67 \text{ x } 10^{-27} \text{ x } (3 \text{ x } 10^8)^2]^2}$$

$$\approx 10^{-10} \text{ Jsm}^{-1}$$

$$\approx \frac{10^{-10}}{1.6 \text{ x } 10^{-19}} = 10^3 \text{ MeV}$$

This is approximately equal to the rest energy of the neutron, *i.e.*, 938 MeV. So, the neutron is inside the nucleus.

(b) Nuclear Spin: Proton-neutron theory solved the problem of nuclear spin as (*p-n*) theory. Since both neutron and proton have the same spin quantum number *i.e.* $\frac{1}{2}$ so the total spin of 'A' nucleons will be a half-integral multiple of *h* depending on 'A' odd or even, respectively.

(c) Nuclear Magnetic Moment: The magnetic moments of proton and neutron are:

$$\mu \, N.P = \frac{eh}{2m_{n,p}} \, Am^2$$

It was observed that magnetic moments of proton and neutron are of the same order as that of nuclear magneton, so it was confirmed that neutrons & proton reside inside the nucleus.

(d) Finite-size: The proton's mass is approximately the same as that of a neutron, and the total number of particles inside the nucleus is equal to the mass number A; hence, the nucleus can easily accommodate it.

(e) β Decay: Proton-neutron hypothesis justifies the emission of β-rays. Since electrons are not present inside the nucleus, it was assumed that electron is created during radioactive disintegration by following reactions,

$$n \rightarrow P + \beta^- +$$

$$P \rightarrow n + \beta^+ + \nu$$

i.e., neutron converting into a proton, an electron, and a new particle called an antineutrino. Likewise, a proton gets converted into a neutron, position and new particle neutrino. Neutrino and antineutrino were the massless particles having spin =12.

(f) Statistics: Proton-neutron theory explained the statistics of all the elements. Both proton and neutron obey Fermi Dirac's statistics. Now a mass number 'A' of elements is the sum of protons and neutrons. Now nucleus with even 'A' will obey Bose-Einstein statistics, and a nucleus with odd A will obey Fermi-Dirac statistics, and these results were in agreement with experimental values.

(g) Isotopic Mass: Proton-neutron theory explained the isotopic masses. Isotopes are defined as the atoms having the same charge number Z (*i.e.*, number of protons) and different mass numbers A (*i.e.*, number of protons (Z) + number of neutrons (N)) So, we can say isotopes have different number of neutrons N only.

1.3. Nuclear Properties

The nuclear properties include the size of the nucleus, spin, party, statistics, magnetic dipole moment, and quadruple moment. All these properties are explained one by one:

1.3.1. Nuclear Size

The first evidence of nuclear size was given by the Rutherford scattering experiment.

According to this experiment, as the α-particle approaches the nucleus (shown in fig. (**3.4**)), the potential energy increases due to the coulomb's repulsive force between the nucleus and α-particle and hence the kinetic energy decreases. A point is reached when kinetic energy is fully converted into P.E. At this point α-particle is turned back. Let the distance between this point and the centre of the nucleus is 'd', also known as the distance of the closest approach.

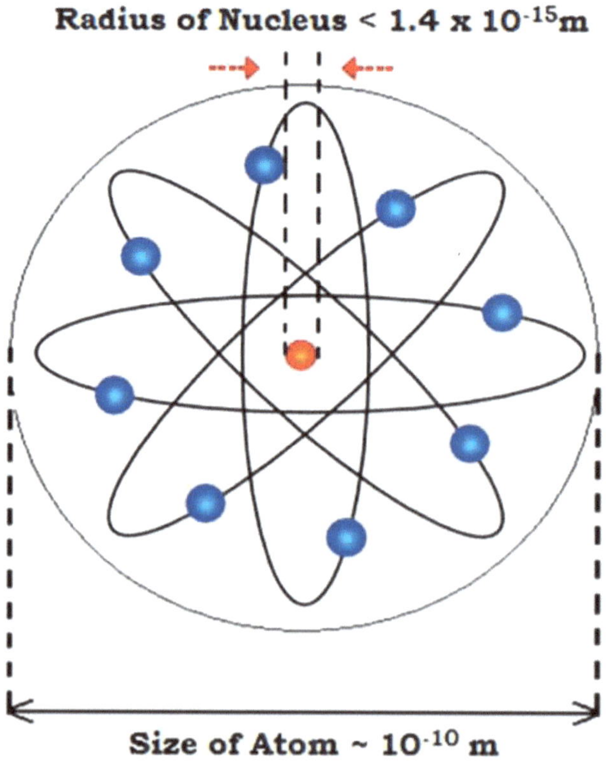

Radius of Nucleus < 1.4 x 10^{-15}m

Size of Atom ~ 10^{-10} m

Fig. (3.4). Nuclear Size.

At point B,

P.E = K.E

Now,

$$P.E = K.E \tag{3.1}$$

$$P.E = \frac{1}{4\pi\epsilon_0}\frac{(Ze)(2e)}{d}$$

Ze → Charge on nucleus

2e → Charge on a particle

d → Distance of closest approach

$$K.E = \frac{1}{2}\,m\upsilon^2$$

υ → Velocity of α particle

Put these values in equation (3.1)

$$\frac{1}{4\pi\epsilon_0}\frac{2Ze^2}{d} = \frac{1}{2}m\upsilon^2$$

$$d = \frac{2\,Ze^2}{4\pi\epsilon_0 E}$$

If the energy of the a particle is 8MeV.

Heavy nucleus was gold (Z= 79) then

$$d = \frac{2 \text{ x } 79 \text{ x } (1.6 \text{ x } 10^{-19}) \text{x } 9 \text{ x } 10^{9}}{8 \text{ x } 1.6 \text{ x } 10^{-13}}$$

$$d = 2.8 \text{ x } 10^{-14}\text{m}$$

In this experiment, the scattering was found to be following the coulomb's law provided the distance between ∝-particle and the heavy nucleus is greater than 10^{-14}m. Coulomb's law could not correctly explain the behaviour of (-particle if this distance is less than 10^{-14}m because a nucleus does not appear as a point charge to the ∝-particle. So, this method failed.

Now different other experiments were given to explain the size of the nucleus. One among them was the Electron scattering experiment. In this experiment, high-energy electrons were scattered from the nuclei. This experiment found that since the nucleus is approximately spherical, its volume is directly proportional to its mass number [11].

$$V \propto A$$

$$\frac{4}{3}\pi r^3 \propto A$$

r → radius of the nucleus

$$r^3 \propto A$$

$$r \propto A^{1/3}$$

$$r = r_0 A^{1/3}$$

r_0 is constant for all nuclei, and its value varies from 1.4 to 1.6 Fermi (1 Fermi = 10^{-15}m) depending upon the experimental method used. The average value of the radius of the nucleus often used is $r_0 = 1.4 \times 10^{-15}$m (See Fig. **3.4**).

1.3.2. Nuclear Spin

A nucleus consists of neutrons and protons. Each proton and neutron possesses an intrinsic angular momentum of magnitude $\frac{h}{2}$ just like an electron. The intrinsic angular momentum is also termed spin. The total angular momentum of nuclei constitutes the spin and orbital angular momenta of the nucleons. The orbital angular momentum is due to the motion of the nucleus about the centre of the nucleus, and the spin angular momentum is due to the spinning motion of the nucleon about its axis, passing through its centre of mass. The magnitude of spin angular momentum constitutes two states, (i) spin parallel to the given direction and (ii) spin anti-parallel to the given direction. So, the component of spin along some direction will be $\frac{h}{2}$ or $\frac{-h}{2}$. Then, the total angular momentum of each nucleon T about a given direction is:

$$i = l \pm s.$$

For a nucleus having more than one nucleon, the resultant angular momentum is:

$$I = L \pm S$$

Where

$$L = \sqrt{l(l+1)h}$$

And

$$S = \sqrt{s(S+1)h}$$

Both proton and neutron are fermions having spin = $\frac{1}{2}$. In the case of even-even nuclei, *i.e.*, an even number of protons and an even number of neutrons, all the protons and neutrons pair off to cancel out one another's spin and orbital angular momenta, *e.g.*, $_2He^4$, $_6C^{12}$*etc.* have zero spins (I = 0). Even odd and odd-even nuclei (*e.g.*, $_4Be^9$, $_6C^{13}$) have half-integral spin. Odd-odd nuclei, an extra neutron and an extra proton yields integral spin.

For experimental verification, Pauli suggested that just like the electrons, atom nuclei also possess orbital and spin angular momentum. Similarly, spin and orbital magnetic moments are associated with angular momenta. The interaction between the two magnetic energy levels is called the hyperfine structure [12].

1.3.3. Parity

A system's parity refers to wave function behavior 'Ψ'.under the inversion of coordinates through the origin. According to quantum mechanics, with every moving particle, a wave is associated that corresponds to the wave function Ψ that depends upon space coordinates (*x, y, z*). Then the probability of finding the particle is $|\Psi|^2 = \Psi\Psi^*$ where Ψ^* is a complex conjugate of Ψ. Since Ψ is a function of (*x, y, z*), if we replace *x to -x, y to -y,* and *z to -z,* then the behaviour of wave function under this inversion of coordinates is called parity, *i.e.*, the wave function is said to have even parity if *y(x, y, z) — y (-x, -y, -z)* and the wave function is said to have odd parity if:

$$\Psi (x, y, z) = -y (-x, -y, -z)$$

In general

$$\Psi\ (-x, -y, -z) = P(y\ (x, y, z)).$$

Where P can be taken as quantum number. P = + 1 given even parity and P = -1 given odd Parity.

There are similar selection rules for P. In the case of a hydrogen-like atom, the parity is related to orbital quantum number *l*, as

Parity P is even for *l* even, and P is odd for *l* odd.

Parity remains conserved during nuclear transformations. Suppose a nucleus is in an excited state and is described by a wave function having odd parity; then, after emitting *y*-rays, the recoiling nucleus plus *y*-rays must continue to have odd parity. The parity of the whole system can be defined as the product of parities of individual particles, *i.e.*, $P = P_1, P_2, P_3 \ldots P_n$, if there are one to n number of particles in a system.

1.3.4. Statistics

The concept of statistics is related to the behavior of many particles. So, to define the properties of assemblies of electrons, protons, neutrons, photons, deuterons and atomic nuclei in general, two kinds of statistics have been developed. These are Bose-Einstein statistics and Fermi Dirac statistics.

(a) Bose-Einstein Statistics: The particles that obey BE statistics are called Bosons. Bosons are particles that have an integral or zero spin. Examples of bosons are Photons, π mesons *etc.* The wave function of a system obeying BE statistics is symmetric, *i.e.*, if the coordinates of the wave function of a system are interchanged, then the wave function describing the new system will remain the same, *i.e.*

$$\Psi(x_1, x_2 \ldots\ldots x_i \ldots\ldots x_j \ldots\ldots x_n) = \Psi\ (x_1, x_2 \ldots . x_j, x_i, x_n)$$

Thus two or more identical can be in the same quantum state now; bi-statistics are obeyed by all the nuclei with an even mass number A.

(b) Fermi Dirac (F-D) Statistics: The particles that obey F-D statistics are called fermions. Fermions are particles that have half spin. Examples of fermions are protons, electrons, neutrons *etc.* The wave function of a system obeying FD statistics is anti-symmetric, *i.e.* if the coordinates of the wave function of the system are interchanged, then the wave function describing the new system will alter in its sign *i.e.*

$$\Psi(x_1, x_2......x_i......x_j......x_n) = \Psi\ (x_1, x_2....x_j, x_i, x_n)$$

This property leads to the important conclusion that only one particle can occupy each completely specified quantum state. Thus, the Pauli Exclusion Principle is obeyed by the particles with FD statistics; hence, we conclude that FD statistics are obeyed by the nuclei having odd mass number A.

1.3.5. Nuclear Magnetic Dipole Moment

'g' is different for different nucleus and g = 2.792 for proton and g = - 1.913 for a neutron,

$$\mu_{proton} = 2.792\ \frac{eh}{2m_p} = 2.792\ \mu_N$$

$$\mu_{proton} = 1.913\ \frac{eh}{2m_p} = 1.913\ \mu_N$$

The magnetic moment value for proton is positive, and that of the neutron is negative because of the complex nature of neutrons. It may be considered that a neutron consists of equally positive and negative charges, and the negative charge is at a long distance from the axis of rotation.

1.3.6. Quadrupole Moment

Quadrupole moment measures the deviation of a nucleus from spherical symmetry. We always assumed that nucleus is spherically symmetric, but this assumption is not always true. So, Q was defined to measure the nucleus's deviation from spherical symmetry and quadrupole moment.

Q is zero (Q = 0) for spherical structure

Q is greater than zero (Q > 0) for prolate,

Q is less than zero (Q < 0) for oblate

Expression:

Let us consider that charge e+ (proton) is not situated at the centre of the nucleus but is located at point Q having rectangular coordinates (*x, y, and z*) as shown in Fig. (**3.5**). The potential point P on the Z-axis due to charge (e+) at Q is,

$$V = \frac{1}{4\pi\epsilon_0}\frac{e}{d}$$

Using cosine formula for triangle OPQ,

$$d^2 = r^2 + r^2 - 2rr\,cos\theta$$

$$d = [r^2 + r^2 - 2rr\,cos\,\theta]^{1/2}$$

$$r\left\{1 + \frac{r^{12}}{r^2} - \frac{2r^1}{r}\,cos\,\theta\right\}^{1/2}$$

$$\frac{1}{d} = \frac{1}{r}\left\{1 + \left(\frac{r^2}{r^2} - \frac{2^r}{r}\,cos\,\theta\right)\right\}^{1/2}$$

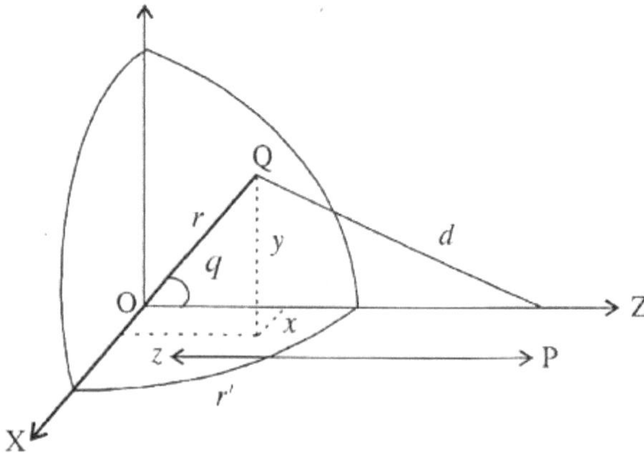

Fig. (3.5). Quadrupole moment of the nucleus: The deviation of a nucleus from spherical symmetry.

Using Binomial theorem,

$$(1 + x)^n = l + nx + \frac{n(n-1)}{2!}x^2 + \frac{n(n-1)(x-2)}{3!}x^3 + \cdots for|x| <$$

$$\frac{1}{d} = \frac{1}{r}\left\{1 + \left(-\frac{1}{2}\right)\left(\frac{r^{12}}{r^2} - \frac{2r^1}{r}\cos\theta\right) + \frac{3}{8}\left(\frac{r^{12}}{r^2} - \frac{2r^1}{r}\cos\theta\right)^2 + \cdots\right\}$$

$$\frac{1}{d}=\frac{1}{r}\left\{1-\frac{r^{12}}{2r^2}+\frac{r^1}{r}\ cos\theta+\frac{3}{4}\left(\frac{r^{14}}{r^4}+\frac{4r^{12}}{r^2}cos^2\theta-\frac{4r^{13}}{r^3}\ cos\theta\right)+..\right\}$$

$$=\frac{1}{r}\left\{1+\frac{r^1}{r}cos\theta-\frac{r^{12}}{2r^2}+\frac{3}{2}\frac{r^{12}}{r^2}\ cos^2\theta+higher\ order\ terms\right\}$$

$$=\frac{1}{r}\left\{1+\frac{r^1}{r}\ cos\theta+\frac{r^{12}}{2r^3}\ (3\ cos^2\theta-1)+\cdots\right\}$$

$$=\left\{\frac{1}{r}+\frac{r^1}{r^2}\ cos\theta+\frac{r^{12}}{2r^3}\ (3\ cos^2\ \theta-1)+\cdots\right\}$$

So

$$V=\frac{e}{4\pi\epsilon_0 r}+\frac{er^1}{4\pi\epsilon_0 r^2}\ cos\theta+\frac{er^{12}}{4\pi\epsilon_0 r^3}\left\{\frac{3}{2}\ cos^2\ \theta-\frac{1}{2}\right\}+\cdots$$

Coefficients of $\frac{1}{r},\frac{1}{r^2},\frac{1}{r^3}$ denote monopole, dipole and quadrupole moment, respectively.

Therefore in c.g.s.quadrupole moment can be given as,

$$Q=er^{12}\left\{\frac{3}{2}\ cos^2\ \theta-\frac{1}{2}\right\}$$

$$Q=er^{12}\left\{{}^{3}/{}_{2}\ cos^2\ \theta-\frac{1}{2}\right\}$$

This expression indicates that the nucleus will only have a finite quadrupole moment when protons are placed asymmetrically.

1.3.7. Nuclear Mass

The mass of an atom is the total of the mass of the nucleus and the orbital electrons. Since the mass of an electron is negligible compared to the mass of the nucleus, it accounts for more than 99.75% of the atom's mass. Experimentally, the nuclear mass can be determined using a mass spectrograph, approximately equal to the mass number A.

The mass of the nucleus is determined in terms of atomic mass unit (*a.m.u*). One atomic mass unit is equal to $\frac{1}{12^{th}}$ of the mass of carbon ${}_6C^{12}$ atom

Since Avogadro number (N) atoms of Carbon has mass = 12g

1 atom of carbon has mass $\frac{12}{N}$g

As,

$$1\ u = \frac{1}{12}\ x\ \text{mass of}\ {}_6C^{12}$$

$$1\ u = \frac{1}{12}\ x\ \frac{12}{N}g$$

$$\frac{1}{N}\ g = \frac{1}{6.023\ x\ 10^{23}g}$$

$$= 1.66\ x\ 10^{-24}g = 1.66\ x\ 10^{-27}kg$$

So,

$$1\ u = 1.66\ x\ 10^{-27}kg$$

In terms of energy

According to Einstein's mass-energy relationship,

$$E = mc^2$$

$$m = 1\ a.m.u = 1.66\ x\ 10^{-27}kg$$

If,

$$C = \text{velocity of light} = 3\ x\ 10^8\ m/s.$$

$$E = 1.66\ x\ 10^{-27}\ x\ 9\ x\ 10^{16}$$

$$E = 1.49\ x\ 10^{-10}\ J$$

Since,

$$1\ \text{electron volt} = 1.6\ x\ 10^{-19}\ J$$

Accordingly,

$$E = \frac{1.49 \times 10^{-10}}{1.6 \times 10^{-19}} \text{ eV}$$

$$E = 931.25 \text{ MeV}$$

$$E \approx 931 \text{ MeV}$$

So,

$$1u = 931 \text{ MeV}$$

Now,

Mass of proton

$$m_p = 1.007\ 276 \text{ u}$$

$$m_p = 1.6725 \times 10^{-27}\text{kg} \approx 938.3 \text{ MeV}$$

$$\text{Mass of neutron } m_n = 1.008\ 665 \text{ u}$$

$$= 1.6748 \times 10^{-27}\text{kg}$$

$$m_n \approx 939.6 \text{ MeV}$$

$$\text{Mass electron} = 0.000549 \text{ u}$$

$$m_e = 9.1 \times 10^{-31}\text{kg}$$

$$m_e \approx 0.51 \text{ MeV}$$

The mass of a neutron is slightly heavier than the mass of a proton.

1.4. Determination of Mass

The study of atomic masses has been of the greatest importance in developing nuclear physics. Different instruments can be used to measure the atomic masses, determine the abundance of isotopes, or produce separated isotopes. One such instrument is the mass spectrometer. A mass spectrometer is an instrument that can measure the masses and relative concentrations of atoms and molecules. In this instrument, an electric or a magnetic field is adjusted to make the spectrum fall on a fixed detecting slit and measured electrically, whereas the instruments in which we use a photographic plate to get a spectrum is called a mass spectrograph.

The exact determination of atomic masses began in 1919 when F. W. Aston invented a mass spectrograph for atomic masses; this mass spectrograph confirmed Soddy's prediction that many stable elements posse isotopes. At about the same time, A.J Dempster developed a spectrometer especially suitable for isotope analysis. It was realized that the O^{16} atom was the most suitable standard and had an atomic mass of 16 units. So, the mass spectroscopes compare the charge to mass (e / m) ratios of concerned cons with the O^{16} atom [13]. Later in 1932, K.T Bain Bridge in the U.S.A. built a mass spectrograph incorporating a velocity filter for accuracy. Later, during 1935 - 1940, several instruments like the Mattauch mass spectrograph, Bainbridge and Jordon mass spectrograph were designed using directional force and velocity focusing (*i.e.*, double-focusing) to get better accuracy and resolving power. Recently several novel time of flight instruments like the mass synchrometer has been invented to get an accurate value of atomic masses. Now let us discuss a few spectrographs:

1.4.1. Bain Bridge's Mass Spectrograph

K.T Bainbridge of U.S.A. built a spectrograph in 1932. In this spectrograph, ions are detected photographically. It is a 180-degree magnetic focusing instrument. The design of the instrument is shown in Fig. (**3.6**).

Ions are formed in ionization chambers at the anode (D), pass through the cathode © and then enter through slits S_1 and S_2, then finally travel between two plates P_1 and $P_{2,}$ kept at positive and negative potentials, respectively. The potential V is applied between plates P_1 and P_2. A magnetic field B is applied perpendicularly, acting at right angles to electrostatic field E.

A particle with a charge q and velocity will only pass through the next slit S_3. The strengths of the two fields are so adjusted that the forces along the ions are equal and opposite. As a result, these ions can pass through the jitter with deflection. In such cases, electromagnetic and electrostatic forces act on the particle at right angles resulting in the Lorentz force field. The condition for this is electromagnetic ($Bq\upsilon$) and electrostatic (qE) forces can be written as,

$$Bq\upsilon) = (qE)$$

• Velocity of particle to pass through S_2 is

$$\upsilon = \frac{E}{B}$$

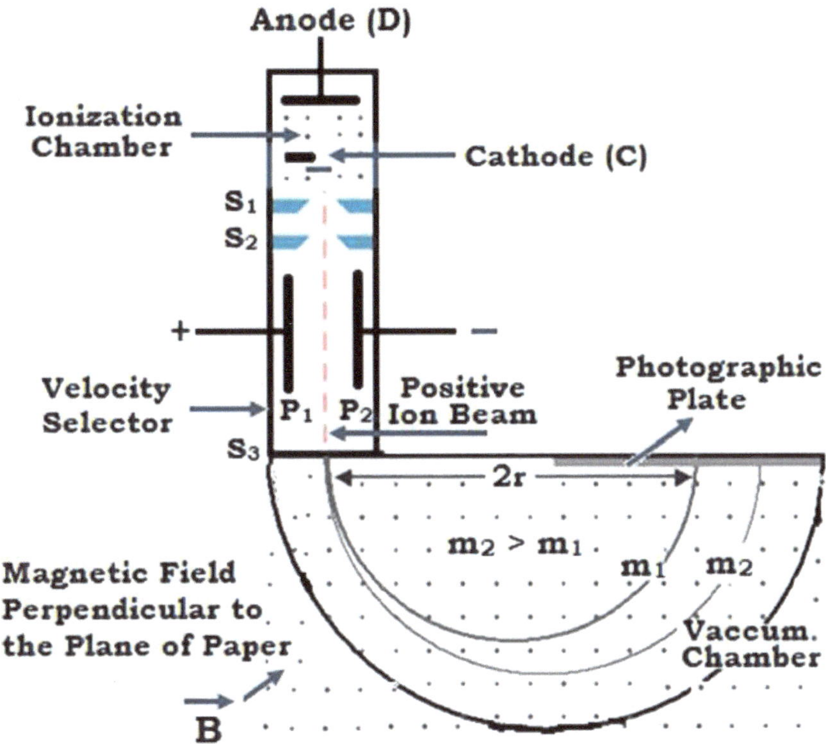

Fig. (3.6). Bain Bridge's Mass spectrograph.

This is a constant, so particles with a certain velocity only can enter the deflection chamber F. The combination of slits and deflection plates is known as a velocity selector.

In the deflector chamber, the ions are affected by the magnetic fields, so they move in a circular path. If the mass of an ion is m, its charge is q, and velocity is υ, then

$$B\upsilon q = \frac{M\upsilon^2}{r}$$

where r is radius of path

$$r = \frac{M\upsilon^2}{Br}$$

the mass of ion 'm' = $\frac{B\upsilon qr}{r\upsilon^2} = \frac{qB^2r}{E}$

The radius of the path is directly proportional to the mass of the ion.

The detection is either a photographic plate or a collector that produces a small current when ions fall on it. The magnetic field may vary, so changing the radii of particle paths are essential so that ions of different masses fall on a fixed collector. The appearance of a photographic plate is shown in Fig. (**3.7**) when a gas containing two isotopes is used.

Fig. (3.7). The appearance of a photographic plate in Bain Bridge's Mass spectrograph.

The wider line for m_1 shows its relatively greater abundance. This analysis method is very accurate and can detect differences in the masses of two ions as small as one part in 10^9.

Bainbridge used this instrument to measure the masses of many lighter atoms. These included the deuterium atom and the atoms involved in the nuclear reaction.

$$^{7}\text{Li} + {}^{1}\text{H} \rightarrow 2({}^{4}\text{He})$$

This work provided the first experimental test of Einstein's mass-energy relationship.

1.4.2. Bain Bridge and Jordon Spectrograph

It is one of the best double-focusing mass spectrographs and is the modified version of the Bainbridge spectrograph. It was invented by K.T Bainbridge and E.B Jordan in 1936.

Before discussing the Bainbridge and Jordan spectrograph, it is important to study its components.

(1) Ion Source: Ions are produced either by heating the suitably coated filaments or by electron bombardment of gases. For example, Ions of materials (Pt, Au, u) *etc.* may be obtained by passing an oscillating spark discharge between the electrodes containing these materials. Also, the positive ions of alkali metals (Li, Na, K *etc.*) may be obtained by heating their nucleus. The ions emerging from the ion source may have a wide range of energy, momentum, and masses and may move in a diverging beam.

(2) Energy Filter: The ions that emerged from the ion source may have different charges and a wide range of energy and masses depending on the isotopes present. The mass spectrograph must be able to focus ions of the same mass at the same place on a photographic plate. Additionally, it must be capable of focusing ions of different energies and different initial directions of motion at the same point; that is why the double forcing mass spectrograph is used.

The ions from the source are given nearly by accelerating them through a high potential difference V_0. The ions emerging from the exit slit should have the same energy, so it acts as an energy filter. The energy filter consists of a pair of cylindrical plates A and B of a large radius of curvature R compared to the spacing of plates d, as shown in Fig. (**3.8**). The potential difference between the plates is V.

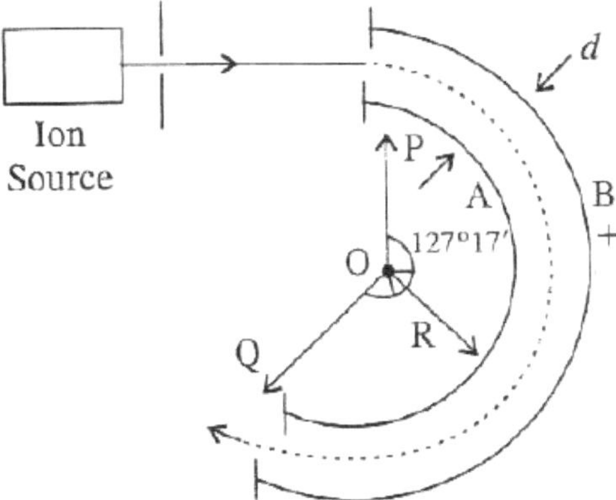

Fig. (3.8). Bain Bridge and Jordon spectrograph.

If the speed of the ion is too high, it will collide with plate B; if the speed is very low, it will collide with plate A. The ions can move along the curvature's central ray of radius if it satisfies the equation.

$$ne\frac{V}{d} = \frac{M\upsilon^2}{R}$$

$$\frac{1}{2}M\upsilon^2 = \frac{neVR}{2d}$$

So the filter will select ions of only definite kinetic energy. It was also shown that direction focusing takes place only if the angular aperture POQ of the cylinder is chosen to be (127° 171). This is a form of direction focusing.

(3) Momentum Filter: The ions emerging from the energy filter will have the same energy, but different ions of different masses will have different masses with different momenta. So to focus ions of the same mass, the beam emerging from the energy filter is passed through a sectorial magnetic field. To get the optimum space focusing on the entrance slit S_1, the apex O of the boundaries of the magnetic field O and the exit slit S_2 of the momentum filter must be on the same straight line as shown in Fig. (**3.9**).

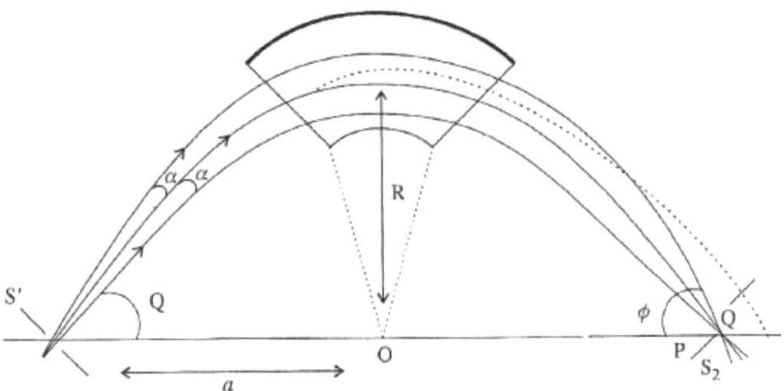

Fig. (3.9). Momentum filter.

The ions entering the magnetic field region follow a circular path with centre O and radius R.

$$\mathrm{Bne}\upsilon = \frac{\mathrm{M}\upsilon^2}{\mathrm{r}}$$

Where B is the flux density of the magnetic field, and it is normal to the plane of particle trajectory. The central ray leaves the magnetic field normal to the field boundary. Particles of the same momentum, including at an angle ± □ to the central rayfocus at point $P_{1,}$ lying closer to O than point Q, where the central ray crosses the line S_1OS_2.

Design of Brain Bridges and Jordan Spectrograph.

The schematic diagram of this instrument is shown in Fig. (**3.10**).

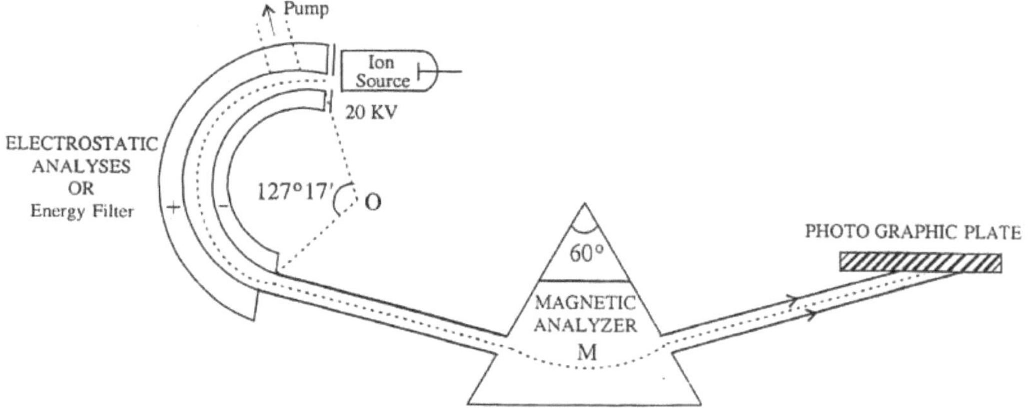

Fig. (3.10). Design of Brain Bridges and Jordan Spectrograph.

In this spectrograph, the energy filter and momentum. The filter is arranged such that the combination produces the best results.

Ions are produced in a high voltage (20 KV) low current discharge tube with broad energy spread into an energy filter maintained at a high vacuum pressure of 10^{-6} mm Hg. The magnetic filter M then analyses the ions emerging from the energy filter into different mass species. Each component is focused on a photographic plate P. Then, the plate is developed, and the positions of lines on the plate are observed with a travelling microscope. After that, the mass resolving power S. Can be calculated as:

$$S = \frac{M}{\Delta M}$$

Where AM corresponds to the width of the spectral line produced on the plate by the ions of mass M. The Bainbridge instrument has a resolving power of the order of 10^4. Also, the dispersion factor ${}^{\Delta M}/{}_{\Delta x}$ can be calculated where Ax is the line spacing.

- **Advantages:** This spectrograph has the advantage that the position of one spectral line can be carefully compared with positions of known mass reference lines. Also, the simultaneous focusing of velocity and direction enables the use of beams diverging in the direction to a greater extent than other instruments. It has high resolving power.

- **Disadvantages:** This instrument has a limited and nonlinear contrast scale of a photographic emulsion. So, it is not desirable for the measurement of isotopic abundance ratios.

1.5. Determination of Charge

The nuclear charge was estimated by Moseley from his work on X-ray spectra. Mosley said a fundamental quantity in the atom increases in regular steps from one element to the next. He identified this quantity as the positive charge on the nucleus and proved that the number of unit charges on the nucleus is the same as the number of places occupied by the element in the periodic table. He called this number the atomic number.

1.5.1. Mosely's Law Explanation

Moseley's law is an empirical law relating to the characteristic x-rays emitted by atoms. The law was revealed and published by the English physicist Henry Moseley in 1913-1914 [14, 15]. Until Moseley's work, 'atomic number' was purely an element's place in the periodic table and was not known to be connected with any calculable physical number [16].

Mosley's law states that the frequency of the corresponding X-ray is proportional to the square of an atomic number of the element of that X-ray.

Mathematically,

$$v_k \propto (Z\text{-}b)^2$$

$$v_k = a\ (Z\text{-}b)^2$$

Where (Z-b) is the effective atomic number of elements entering X-rays, 'a' is a constant, b is the screening constant, and v_K is the frequency of any line in the K-series. For $K_{2,}$ line A was found to be equal to 34 RC, where R is the Rydberg constant and c is the velocity of light. For K_d, b is equal to 1 therefore,

$$v_{(k_\propto)} = \frac{3}{4}\,\text{RC}\ (Z\text{-}1)^2$$

Mosley's law:

Mosley measured the frequencies of characteristic X-ray spectrum lines of about 40 different elements. He noted the plot between atomic number Z and the square root of frequency would be a straight line, as shown in Fig. (**3.11**).

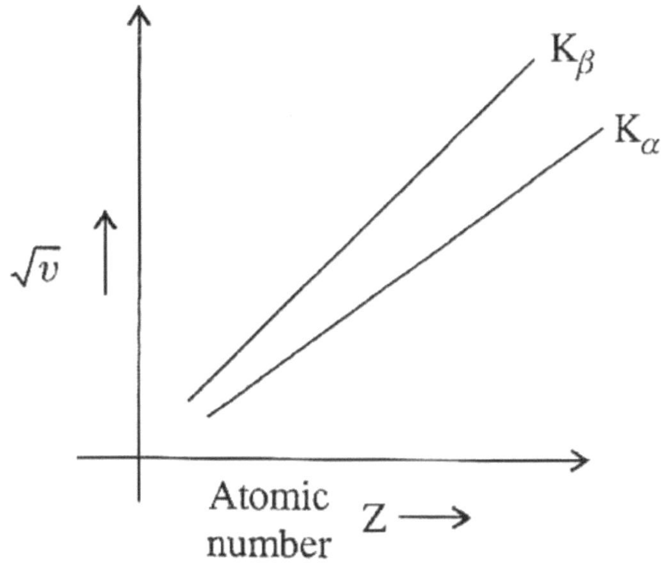

Fig. (3.11). Variation of atomic number (Z) *versus* square root of the frequency of X-ray.

This plot is known as Mosley's plot; the exact Mosley's law is:

$$\frac{1}{\lambda} = \mathrm{RC}(\mathrm{Z} - \mathrm{b})^2 \left(\frac{1}{r_2^2} - \frac{1}{n_1^2}\right)$$

Where n_1 and n_2 are the principal quantum numbers of the energy levels between which the transition takes place and λ is the wavelength.

Proof: According to Bohr Theory, the energy of an electron in the nth orbit.

So for levels n_1 and n_2, the corresponding energy of an electron is,

$$E_{n1} = -\frac{2\pi^2 me^4 Z^2}{n_1^2 \lambda^2}$$

$$E_{n2} = -\frac{2\pi^2 me^4 Z^2}{n_2^2 \lambda^2}$$

Considering the screening effect, the Z is reduced to (Z-b)

$$E_{n1} - E_{n2} = \frac{2\pi^2 me^4}{h^2}\left(\frac{(Z-b_2)^2}{n_2^2} - \frac{(Z-b_1)^2}{n_1^2}\right)$$

$$= \frac{2\pi^2 me^4}{h^2}(Z-b_2)^2\left\{\frac{1}{n_g^2} - \frac{(Z-b_1)^2}{(Z-b_2)^2}\cdot\frac{1}{n_1^2}\right\}$$

When Z is high,

$$b_1 \approx b_2 \approx b$$

$$\therefore \quad E_{n1} - E_{n2} = hv = \frac{2\pi^2 me^4}{h^2}(z-b)^2\left\{\frac{1}{b_2^2} - \frac{1}{n_1^2}\right\}$$

For Kα line $n_1 = 2$. $n_2 = 1$ and b = 1

$$n(k_\alpha) = \frac{2\pi^2 me^4}{h^3}(z-1)^2\left\{\frac{1}{1} - \frac{1}{4}\right\}$$

$$n(k_\alpha) = \frac{3}{4}\frac{2\pi^2 me^4}{h^2}(z-1)^2$$

$$= \frac{3}{4}\,RC\,(z-1)^2$$

Where,

$$R = \frac{2\pi^2 me^4}{ch^3}$$

$$\therefore \quad n(k_\alpha) \propto (z-1)^2$$

This is Moseley's law.

1.5.2. Applications of Mosley's Law

1. Moseley's law helps determine the atomic number of rare earth elements, thereby fixing their position in the periodic table.

2. It also corrected the periodic table by the discovery of new elements such as technetium (^{43}Tc), hafnium (^{72}Hf), rhenium (^{75}Re), *etc.*

3. It showed the positive charge on a nuclear or atomic number as the determining factor of the periodic table rather than atomic weights.

4. Moseley's law can remove any inconsistency in order of the elements in the periodic table by positioning the elements according to the atomic numbers but not according to the atomic weights.

REFERENCES

[1] The Rutherford Experiment. *Archived from the original* Rutgers University., 2001.

[2] D. Harper, "Nucleus", *Online Etymology Dictionary*.

[3] G.N. Lewis, "The Atom and the Molecule", *J. Am. Chem. Soc.,* vol. 38, no. 4, pp. 762-785, 1916. [http://dx.doi.org/10.1021/ja02261a002]

[4] Sitenko, A.G. & Tartakovskiĭ, V.K. Theory of Nucleus: Nuclear Structure and Nuclear Interaction. Kluwer Academic. p. 3. ISBN 978-0-7923-4423-0. 1997. [http://dx.doi.org/10.1007/978-94-011-5772-8]

[5] Battersby, and Stephen, "Pear-shaped nucleus boosts search for new physics", *Nature.,* 2013. [http://dx.doi.org/10.1038/nature.2013.12952]

[6] Srednicki, M.A. Quantum Field Theory. Cambridge University Press. pp. 522–523. ISBN 978-0-5-1-86449-7. 2007. [http://dx.doi.org/10.1017/CBO9780511813917]

[7] Basdevant, J.-L.; Rich, J. & Spiro, M. Fundamentals in Nuclear Physics. Springer. p. 155. ISBN 978-0-387-01672-6. 2005.

[8] R. Machleidt, and D.R Entem, Chiral effective field theory and nuclear forces". *Physics Reports.*, 503 (1): 1–75, 2011. [http://dx.doi.org/10.1016/j.physrep.2011.02.001]

[9] L.P. Gaffney, P.A. Butler, M. Scheck, A.B. Hayes, and F. Wenander, "Studies of pear-shaped nuclei using accelerated radioactive beams" (PDF)", *Nature.,* vol. 497, no. 7448, pp. 199-204, 2013. [http://dx.doi.org/10.1038/nature12073]

[10] a) R. Serway, C. Vuille, and J. Faughn, *College Physics.* 8th 2009. b) N.D. Cook, Belmont, CA: Brooks/Cole, Cengage Learning. *Models of the Atomic Nucleus* 2nd. Springer, 2010, p. 915.

[11] Krane, K.S. Introductory Nuclear Physics. Wiley-VCH. ISBN 978-0-471-80553-3. 1987.

[12] B. Fernandez, and G. Ripka, Nuclear Theory after the Discovery of the Neutron.*Unravelling the Mystery of the Atomic Nucleus: A Sixty Year Journey 1896 — 1956.* Springer, 2012, p. 263.

[13] A.I. Miller, *Early Quantum Electrodynamics: A Sourcebook.* Cambridge University Press: Cambridge, 1995, pp. 84-88.

[14] Moseley, and G.J. Henry, Smithsonian Libraries. "The High-Frequency Spectra of the Elements".*The London, Edinburgh and Dublin Philosophical Magazine and Journal of Science.* vol. 26. Taylor & Francis: London- Edinburgh: London, 1913, pp. 1024-1034.

[15] H.G.J. Moseley, "LXXX. The high-frequency spectra of the elements. Part II", *Lond. Edinb. Dublin Philos. Mag. J. Sci.,* vol. 27, no. 160, pp. 703-713, 1914. [http://dx.doi.org/10.1080/14786440408635141]

[16] J. Mehra, and H. Rechenberg, *The historical development of quantum theory.* vol. 1. Springer-Verlag: New York, 1982, pp. 193-196. [http://dx.doi.org/10.1007/978-1-4612-5783-7]

CHAPTER 4

Particle Detectors

Abstract: Many of the particle detectors developed are gaseous based on ionization. The semiconductor and scintillation detectors are the most typical and cast-off, but other, entirely diverse ideologies have also been functional, like Cerenkov light and transition radiation. Recent detectors in particle physics pool numerous elements in layers much like an onion. The present chapter is aimed to deliberate the significant aspects of G. M. Counter (Geiger-Muller Counters), Wilson cloud chamber, Scintillation counter and a semiconductor detector.

Keywords: Counter, Detector, G. M, Particle, Wilson Cloud.

1. INTRODUCTION TO PARTICLE DETECTORS

A particle detector is an instrument that is used to identify the ionizing particles and their track produced by reactions in a particle accelerator, nuclear decay and cosmic radiation. Detectors can also measure particle energy and other characteristics such as particle type, charge, momentum, spin *etc.* They can also be used to record the particle's presence and are also known as a radiation detector. In investigational and practical nuclear physics, particle physics and nuclear engineering, a particle detector, also recognized as a radiation detector [1], and is a systematic structure recycled to detect, track, and/or classify ionizing particles, such as those produced by nuclear decay, cosmic radiation, or responses in a particle accelerator. Detectors can quantity the particle energy and supplementary characteristics such as particle type, spin, momentum, and charge in accretion to simply classify the particle's existence [2].

Detectors aimed for contemporary accelerators are massive in magnitude and price. The term counter is regularly used as an alternative of detector when the detector computes the particles but does not determine their energy or ionization. Particle detectors can also characteristically track -energy photons or detectable photons (light), known as ionizing radiation. If their primary purpose is radiation measurement, they are entitled radiation detectors, but as photons are also (massless) particles, the term particle detector is quite precise. The following

Ritesh Kohale, Sanjay J. Dhoble & Vibha Chopra

types of particle detectors are commonly used in particle and nuclear physics for radiation protection in the nuclear, medical and environmental fields [3].

1.1. G. M. Counter (Geiger-Muller Counters)

Germany, in 1928, Geiger and Muller settled a 'Particle detector' for determining 'ionizing radiation.' They titled it 'Geiger Muller Counter.' It has been one of the supreme and extensively used nuclear detectors in the initial generations of nuclear physics. It is a gas-occupied counter that functions in the Geiger region. Basically, the G.M. counter is a radiation detector.

A Geiger counter is an apparatus cast-off for spotting and calculating ionizing radiation, and it is comprehensively used in solicitations such as radiological protection, radiation dosimetry, investigational physics, and nuclear engineering. It notices ionizing radiation such as alpha particles, beta particles, and gamma rays using the ionization effect produced in a Geiger–Müller tube, which provides its name to the instrument [4]. It is conceivably one of the world's best-known radiation detection devices in wide and noticeable use as a hand-held radiation investigation device.

The unique detection principle was recognized in 1908 at the University of Manchester [5], but it was not in anticipation of the development of the Geiger–Müller tube in 1928 that the Geiger counter could be manufactured as a practical instrument. Since then, it has been widespread due to its strong sensing component and low cost. Nevertheless, there are restrictions in calculating high radiation rates and the energy of incident radiation [6].

1.1.1. Principle, Construction and Working

Principle

Radioactive rays ionize the gas they pass and yield a small number of ions. In definite conditions, when the applied voltage is adequate, these ions yield a secondary avalanche, and a small voltage drop is recorded through the load. This voltage is intensified so the counter can record it [7]. The Schematics of G. M. Counter are shown in Fig. (**4.1**).

Construction

- The G. M. tube entails a high-pitched metal cylinder performing as a cathode.
- A wire of tungsten drives from side to side to the axis of the tube, which performs as an anode.
- Cathode (-) and anode (+) is disjointed by the insulator (ebonite plugs).

- Together the cathode and anode are coupled with a high voltage DC battery (1000 – 2000 Volts), and load R_L is connected in series with maximum resistance.
- At the one end, a skinny mica window is settled to permit the admittance of radiation in thtube-initiated radioactive source.
- The tube is emptied (vacuum) and then packed with a gaseous mixture *i.e.*, 90% Argon and 10% ethyl alcohol vapours at preferred low pressure.

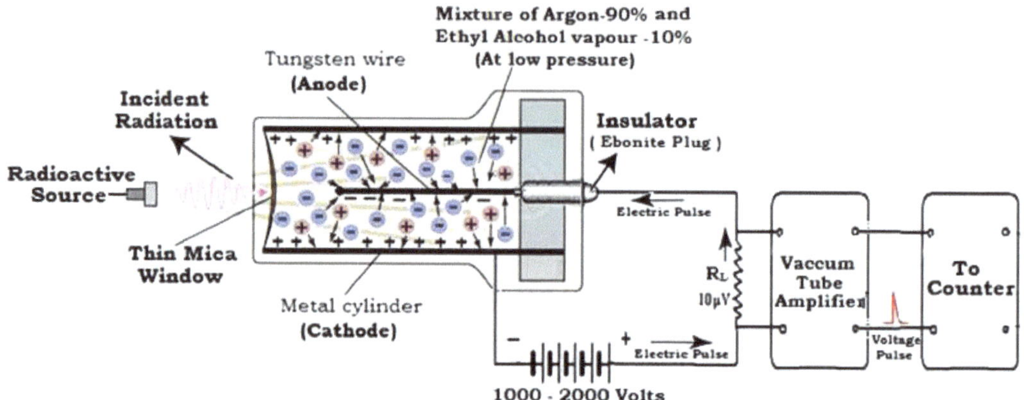

Fig. (4.1). Schematics of G. M. Counter.

The simplified G.M. counter circuit and block diagram are as shown in Fig. (**4.2**).

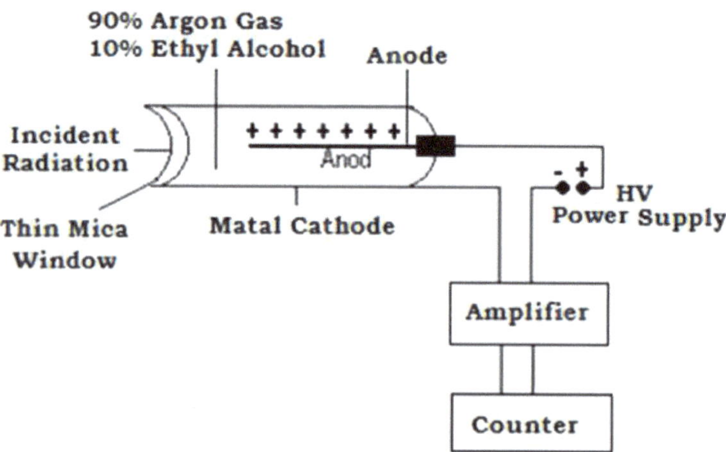

Fig. (4.2). Block diagram of G. M. Counter.

Working

- A dc potential nearby 1200 volt is put on amongst the cathode and anode.
- When the radiation arrives in the GM tube over the mica window, it ionizes a small number of argon atoms.
- If the practically supplied voltage is resilient and adequate, these ions yield a subordinate avalanche that originates an electric current pulse.
- The current from end to end with a load resistance R_L yields a voltage pulse of the order of 10 μV.
- This voltage is enhanced up to 5 -15 Volt by vacuum tube amplifiers and is provided to the counter.
- As each arriving radiation appears in the form of pulse, the number of radiation is counted [8].

1.1.2. Characteristics of G M Counter

Fig. (**4.3**) shows the characteristic plot of count/min as a function of voltage.

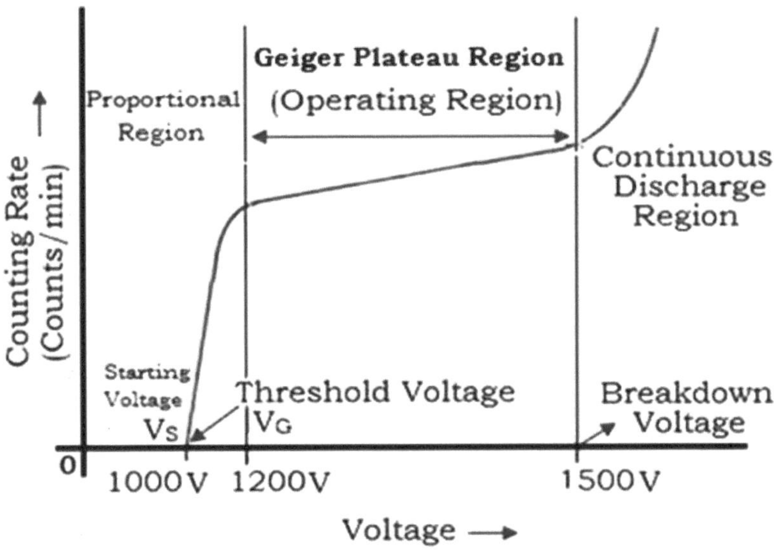

Fig. (4.3). Plot of count/min as a function of voltage.

- Aimed at a voltage less than 1000 Volt, there is no release of energy and, therefore, no counts.
- Stuck among 1000 – 1200 Volts, the amount of pulses calculation rises rectilinearly with the operative voltage.

- This region is characterized as a proportional region
- Beyond 1200 Volt up to 1500 Volt, the count rate displays minimum deviancy and remains essentially continuous; this region is called the plateau region or Geiger region or Operating region.
- If the voltage is applied above 1500 volt, an infinite discharge will proceed, and the count rate intensifies briskly due to liberation of Argon gas, which is surplus.
- A graph is plotted between the counting rate (counts/min) and applied voltage, known as counting characteristics. The counting starts at an approximately precise voltage, Vg, known as threshold voltage, and then displays a sharp increase with an increase in applied voltage. It then settles into an almost flat region [9]. Fig. (**4.3**) indicates the plot of count/min as a function of voltage.

1.1.3. Plateau Characteristics of G. M. Counter

In the G. M. Counter, the radioactive source is kept at a particular distance from the tube. The applied voltage is increased gradually, and the number of particles detected in certain time intervals by G. M. Counter is noted.

Above threshold voltage $V_{G,}$ around 200 V- 400 V, the counting rate remains nearly constant. The range of voltage over which the counting rate remains constant is called **Geiger Plateau.** Beyond the Plateau region, the consistent discharge starts, and counting becomes impossible. Consequently, the G. M. counter must be operated in the Plateau region.

1.1.4. Important Parameters Which Decide the Quality of Functioning of GM Tubes [10]

i. Quenching

It is the progress in the process to sidestep the endless discharge of energy. Vapours of ethyl alcohol. It does self-quenching since its ionization energy is not considerably considered enough compared to the the ionization energy of the Argon atom.

ii. Counting Rate

The G M Counter can figure out around 5000 particles/sec. The counting rate is subjected to G M Counter's death and recovery time.

iii. Dead Time (t)

In the counter, the progressively moving positive argon-ion takes 100 sec to reach the cathode. If the simultaneous (secondary) radiation comes to the tube for the period of this time, it will not be recorded; this time is termed the dead time of the counter and is symbolized by 't.'

iv. Recovery Time

After a dead time, the tube takes an extra 100 μsec to progress and recover the preliminary functioning situation. This interval of time is known as the recovery time of the counter.

v. Resolving (Paralysis) Time

The summation of dead and recovery time is known as resolving time or paralysis time, which is 200 sec. The tube can return to the second radiation later after 200 sec.

Fig. (**4.4**) shows the connection between dead time, recovery time and resolving time.

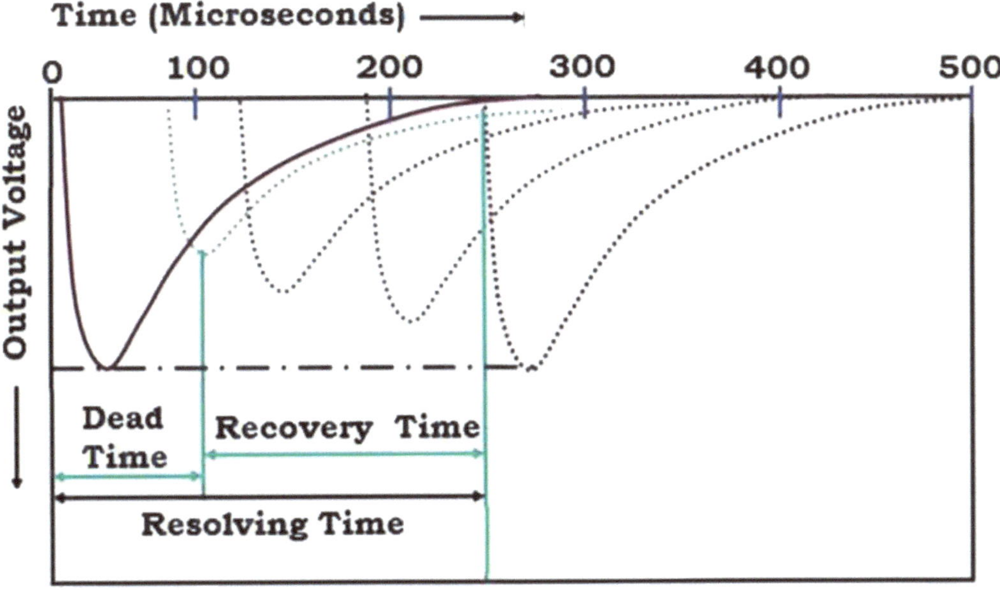

Fig. (4.4). (a): Connection between dead time, recovery time and resolving time of G. M. Counter.

vi. True Count Rate (n)

If the death time of the counter is't' and the count which is observed (measured) by the counter is n_{obs} then the true count rate n,

$$n = \frac{n_{obs}}{1 - n_{obs} \, x \, t}$$

vii. Efficiency of G. M. Counter (η)

It is the ratio of the number of counts detected per second (N_{obs}) to that of the total no of ionizing particles incoming per second (N) in the counter.

$$\therefore \text{Efficiency,} \quad \eta = \frac{N_{obs}}{N}$$

Example: If 1000 particles are arriving at the counter per second and 800 particles are recorded at the same time by the counter, then the efficiency of the counter is given by:

$$\therefore \text{Efficiency,} \quad \eta = \frac{800}{1000} x100 = 80\%$$

1.1.5. Main Features of a Geiger-Muller Counter

- Constant output pulse size, independent of initial ionization.
- A long intensive time to allow entry of each particle.
- The insensitive time is usually made definite by decreasing the applied voltage with the help of a suitable design of an external electrical circuit; it is then called the 'Paralysis Time.'
- It is sensitive to the production of even a single ion pair.
- It can detect V and other cosmic rays.
- It cannot detect uncharged particles like neutrons.

1.1.6. Advantages and Disadvantages of G. M. Counter

Advantages

- It is very accurate and efficient for detecting nuclear particles like protons, α and β particles.

- The pulse height of the plot is independent of the energy and nature of the incident ionizing particle.

- The pulse height is constant over a plateau width.

- With the minimum voltage amplification, sufficiently high pulse height can be achieved.

Disadvantages

• G.M. tube has a limited lifetime.

• It cannot detect neutral and uncharged particles like neutrons.

• The dead time of the G. M. counter is very high.

• γ-radiation detection shows very low efficiency.

1.2. Wilson Cloud Chamber

It is an apparatus recycled to reveal subatomic particles' track (pathway). In Wilson Cloud Chamber, tracks of subatomic particles or ionized particles can take pictures of particles. The cloud chamber, also known as the Wilson chamber, is used to perceive particles of ionizing radiation. In its supreme fundamental arrangement, a cloud chamber consists of vacuum-packed surroundings enclosing a supercooled, supersaturated water or alcohol vapor. When an alpha particle or beta particle interacts with the mixture, it ionizes it.

Cloud chambers played a noticeable part in elemental as well as particle physical science on or after the 1920s to the 1950s, up until the establishment of the bubble chamber. In specific, the findings of the positron in 1932 and the muon in 1936, together with Carl Anderson (endowed a Nobel Prize in Physics in 1936), recycled cloud chambers. In 1947 revolution of the kaon by Clifford Charles Butler and George Rochester was made using a cloud chamber as the detector [11]. In the collective case, cosmic rays were the foundation of ionizing radiation.

Charles Thomson Rees Wilson (1869–1959), a Scottish physicist, is recognized with determination in the cloud chamber. Motivated by detections of the Brocken spectre while functioning on the summit of Ben Nevis in 1894, he initiated advanced enlargement chambers for reviewing cloud formation and photosensitive occurrences in the humid air. Very quickly, he exposed that ions may act as centers for water dew drop creation in such chambers. He followed the submission of this innovation and improved the first cloud chamber in 1911. In Wilson's unique chamber, the air inside the closed device was soaked with water vapor; then a diaphragm was used to swell the air inside the chamber, called adiabatic expansion, cooling the air and starting to condense water vapor. Therefore the term expansion cloud chamber is used. When an ionizing particle allocates from the chamber, water vapor condenses on the subsequent ions, and the path of the particle is apparent in the vapor smog. Wilson, along with Arthur Compton, was acknowledged with the Nobel Prize in Physics in 1927 for his work on the cloud chamber [12]. This kind of chamber is also called a pulsated chamber because the situations for the process are not persistently conserved. Further advances were completed by Patrick Blackett, who employed a steady mechanism to expand and compress the chamber very promptly, building the chamber profound to particles with an elevated frequency of numerous times a second. A vivid film was used to record the pictures. In 1936, Sir Alexander Langsdorf established the diffusion cloud chamber [13]. This chamber was not the same as the expansion cloud chamber as it was noticeably sensitized to radiation, and the lowermost portion of the chamber must be air-conditioned reasonably to a low temperature, commonly colder than −26 °C (−15 °F). As an alternative to water vapor, alcohol is used because of its lower freezing point. The alcohol used in them is commonly isopropyl alcohol or methylated spirit.

The only original cloud chamber used by C.T.R. Wilson to detect the traces of subatomic particles is shown in Fig. (**4.5**). According to Wilson, the material price for the experiment was only around five British pounds. The historic cloud chamber is in the Cavendish Laboratory at Cambridge University [14]. Fig. (**4.5**) reveals the historic cloud chamber located in the Cavendish Laboratory.

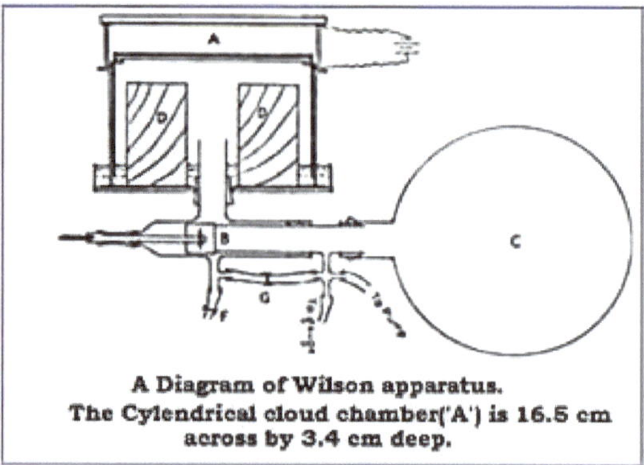

Fig. (4.5). The historic cloud chamber located in the Cavendish Laboratory, Adapted from [15].

1.2.1. Principle, Construction and Working Wilson Cloud Chamber

Principle

When a particle is passed through the supersaturated vapors, droplets are formed on the line due to ionization along the track and the particle is detected [16].

Construction

- It consists of a closed cylindrical chamber with transparent glass at the top and a movable piston at the bottom.

- The cylinder is provided with a glass window on the sides near the top.
- Inside the cylinder, a liquid of low boiling point is placed.
- The piston can be moved up or down.
- The whole system is air-tight.
- A strong light source issued to illuminate the chamber while the photograph is taken by the camera as shown in Fig. (**4.6**).

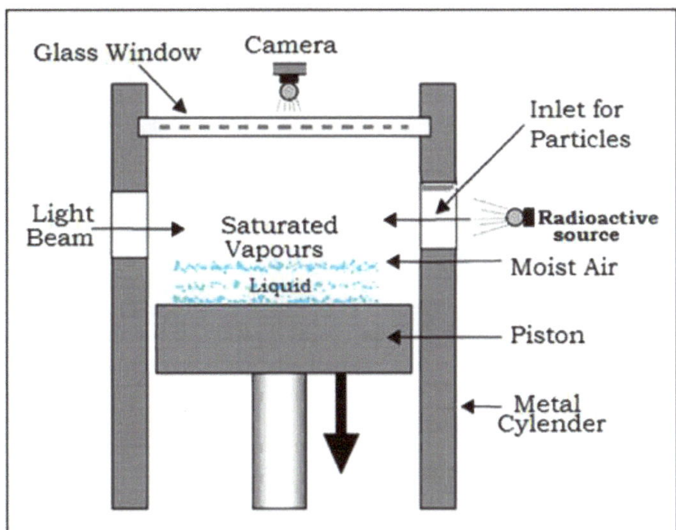

Fig. (4.6). Schematics of Wilson cloud chamber.

Working

- Some volatile liquid having low boiling point (methanol CH_3OH or ethanol C_2H_5OH) is poured on the inner surface of the chamber.
- The piston first is moved slowly up so that the air inside the chamber is cleaned and then moved down, so that the internal pressure is dropped and the air gets vapors of the liquid and becomes supersaturated and a fog is observed in the chamber.
- At the right moment particles are allowed to enter into the chamber and a powerful and intense beam of light is used to illuminate the track of the particles and photos are taken by the sensitive camera
- If a strong electric or magnetic field is applied to the particles (charged), then their path is altered.
- By the study of the path's length, thickness, continuity or discontinuity and the influence of magnetic field (curve). *i.e.*, geometry, the e/m ratio can be calculated and hence the particle is detected.

1.2.2. Paths of the Particles Traced by Wilson Cloud Chamber

The resulting ions act as condensation nuclei, around which a fog will form (because the mixture is on the point of condensation). The high energies of alpha and beta particles mean that a trail is left due to many ions being produced along the path of the charged particle. These tracks have distinctive shapes, as shown in Fig. (**4.7**).

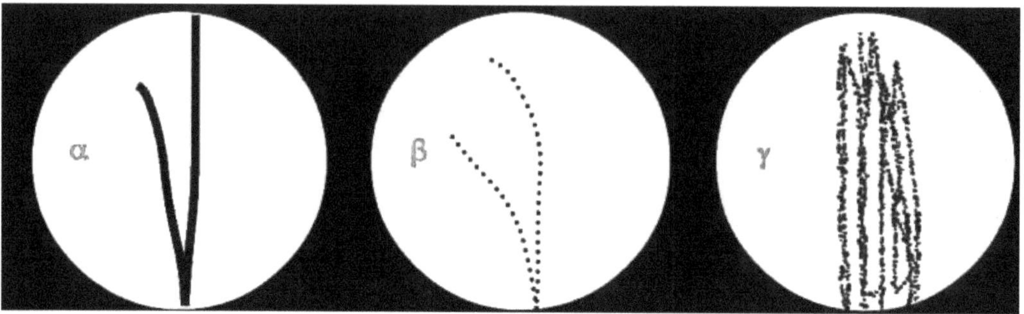

Fig. (4.7). Paths of particles.

- An alpha particle's track is broad and straight,
- An electron's is thinner and shows more evidence of deflection by collisions).
- When any uniform magnetic field is applied across the cloud chamber, positively and negatively charged particles will curve in opposite directions, according to the Lorenz force law with two particles of opposite charge.

1.2.3. Applications of Wilson Cloud Chamber

- It is widely used in the study of cosmic rays.
- It can play a vital role in the investigations of novel elementary particles.
- It is useful to study the properties of particles like mesons, bosons and positrons.

1.3. Scintillation Counter

A scintillation counter is an instrument for identifying and calculating ionizing radiation by using the excitation effect of incident radiation on a scintillating material and sensing the resulting light pulses. It involves a scintillator that produces photons in reaction to incident radiation, a sensitive photomultiplier tube (PMT) Photo detector, a charge-coupled device (CCD) camera, or a photodiode), which transforms the light into an electrical signal and electronics to process this signal. Scintillation counters are extensively recycled in radiation protection, an assay of radioactive materials and physics research because they can be made reasonably yet with good quantum efficiency and can measure both the concentration and the energy of incident radiation.

The contemporary scintillation counter was developed in 1944 by Sir Samuel Curran [17], although he was working on some other project. There was a necessity to count the radiation from minor extents of uranium, and his discovery was to utilize one of the freshly- accessible and enormously complex photomultiplier tubes prepared by the Radio Corporation of the USA to count the sparks of light and to acquire their careful clarifications beginning with a scintillator exposed to radiation. This scintillator was assembled upon the effort of previous investigators such as Antoine Henri Becquerel, who exposed radioactivity and employed it on the phosphorescence of uranium salts in 1896. Beforehand scintillation dealings had to be meticulously perceived by using a spinthariscope, a self-effacing microscope, to perceive light flares in the scintillator [18].

1.3.1. Principle of Operation

As soon as an ionizing particle is allowed to pass into the scintillation substantially, particles are agitated alongside a trajectory. The track for stimulating charged elements and particles in the track of the particle itself. For gamma rays (uncharged particles), their energy is transformed into an energetic electron *via* either the photoelectric effect, Compton scattering or pair production. The working principal of scintillation counter in which incident high energy photon striking a scintillating crystal, exciting the discharge of low-energy photons which are then transmuted into photoelectrons and replicated in the photomultiplier is sketched in Fig. (**4.8**).

The chemistry of atomic de-excitation in the scintillator yields a swarm of low-energy photons, characteristically near the blue region of the visible spectrum. The magnitude of such photons is interactive with the energy established by the ionizing particle. These resultant photons can be focused on the photocathode of a photomultiplier tube (PMT) which discharges a single electron for each incoming photon as a result of the photoelectric effect. This assembly of principal electrons is electrostatically augmented and concentrated by an electrical potential so that they attack the initial dynode of the tube. The influence of a single electron on the dynode discharges a sufficient amount of subordinate electrons, which are fast-tracked to smash another dynode. The influence of each succeeding dynode produces additional electrons, so there is a current magnifying effect at the respective dynode point. Each phase is at a higher potential than the preceding to deliver the accelerating field. The output signal obtained at the anode is a quantifiable pulse for each group of photons from an initial ionizing incident in the scintillator that is grasped at the photocathode and transfers information regarding the energy of the initial radiation. When it is transported to a charge amplifier that assimilates the energy statistics, the pulses produced in this process are compared to the particle energy responsible for electrification of the scintillator. The rate of generation of such pulses concerning time confirms the concentration as well as the intensity of the radiation. Approximately in most applications, individual pulses are not summed up, but merely the typical current at the anode is considered to identify the degree of radiation intensity. The scintillator should be secured from all external light so that such photons do not marsh the ionization actions produced by arriving emission. To accomplish such experimental conditions, a thin smoky foil, such as aluminized mylar, is frequently used; nevertheless, it needs to have an adequately low mass to decrease excessive fall of the incident radiation that is to be dignified and counted.

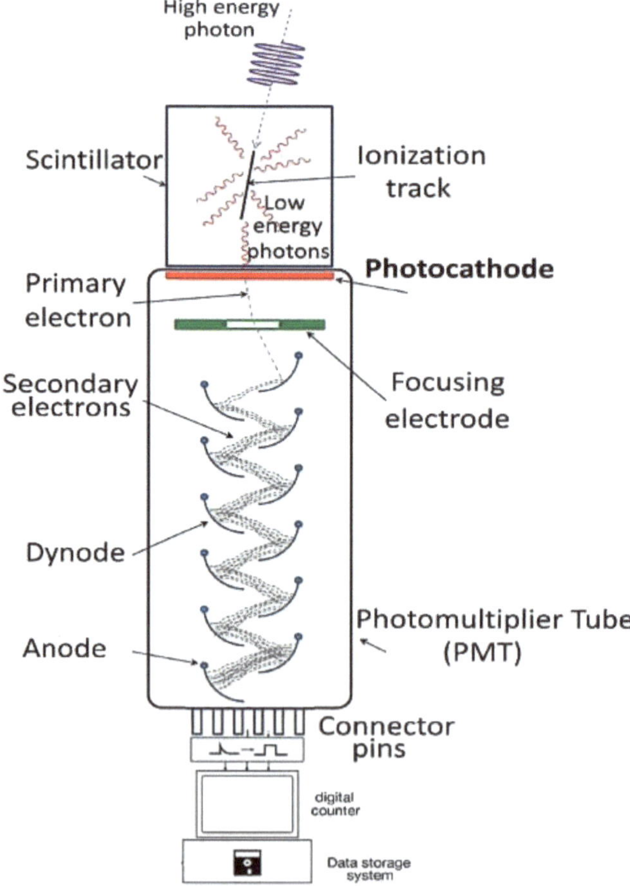

Fig. (4.8). Schematic showing Scintillation Counter.

1.3.2. Efficiencies of Scintillation Detector

a. Neutron

In the case of neutron detectors, extraordinary effectiveness is enlarged through the use of scintillating materials rich in hydrogen that scatter neutrons capably. Liquid scintillation counters are a competent and concrete means of quantifying beta radiation.

b. Gamma

The quantum efficiency per unit volume of a gamma-ray detector may be influenced by the concentration of electrons inside the detector, and definite scintillating constituents, for instance, bismuth germanate and sodium iodide accomplish extraordinary electron concentrations as a consequence of the high atomic numbers of which they are comprised. On the other hand, hyperpure germanium semiconducting sensors have improved fundamental energy perseverance than scintillators and are favoured reasonable gamma-ray spectrometry.

1.3.3. Applications of Scintillation Detector

A number of products have been brought together in the marketplace employing scintillation counters for recognition of actually hazardous gamma-emitting resources during transportation. These contain scintillation counters aimed at contamination monitoring of nuclear waste, border security, cargo stations, weigh bridge applications, ports and scrap metal yards [19]. There are alternatives of scintillation counters attached to helicopters and trucks for rapid reaction in case of safety circumstances due to radioactive waste or muted bombs [20]. Hand-seized units are also ordinarily used.

Scintillation probe being used to measure external radioactive impurity as well as to count the degree of radiation in a range of applications together with hand whispered radiation investigation meters, ecological and personnel checking for radioactive infection, radiometric inspect, medical imaging, nuclear plant security and nuclear retreat [21].

1.4. Semiconductor Detector

A semiconductor detector in ionizing radiation detection physics is a method that typically applies silicon or germanium semiconductors to measure the degree of influence of incident photons or stimulating particles. Semiconductor detectors treasure wide-ranging solicitations such as particle detectors, radiation protection and gamma and X-ray spectrometry.

1.4.1. Principle of Operation

The ionizing radiation is signified by the figure of charge transporters established and allowed to pass in the detector material sandwiched in the middle of two probes by the radiation. Ionizing radiation generates free electrons and holes. The energy of the radiation is proportionate to the number of electron-hole pairs generated in the semiconductor. Consequently, a number of electrons are conducted from the valence band to the conduction band, and an equivalent number of holes are produced in the valence band. Under the effect of an electric field, electrons and holes get mobilized towards the electrodes, where they upshot in a pulse and can be easily measured in an external circuit. The holes travel in the reverse track and can also be measured. As the amount of energy essential to generate an electron-hole pair is allowed and does not depend upon the energy of the incoming radiation, determining the number of electron-hole pairs permits the intensity of the incident radiation to be determined [22]. The energy required to produce electron-hole pairs are very little compared to the energy essential to return and join the ions in a gas detector. Therefore, in semiconductor detectors, the algebraic dissimilarity of the pulse height is reduced, and the persistence of energy is improved. As the electrons have enough kinetic energy and move fast, the perseverance of time is also excellent, dependent upon escalation time [23]. Matched with gaseous ionization detectors, the compactness of a semiconductor detector is extraordinary, and charged particles of utmost energy can contribute their energy in a semiconductor of practically insignificant magnitudes. A schematic diagram of the semiconductor detector is shown in Fig. (**4.9**).

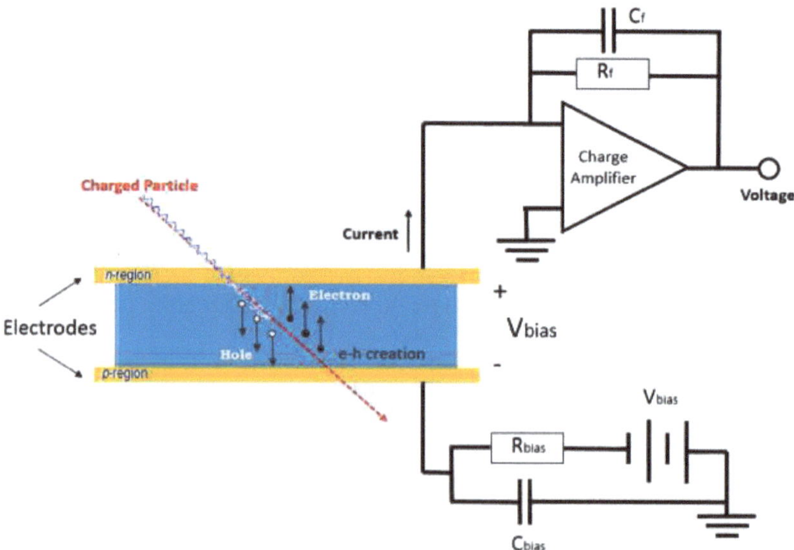

Fig. (4.9). Schematic diagram of Semiconductor detector.

1.4.2. Types of Semiconductor Detectors

a. Silicon Detectors

A PHENIX detector is a forward Silicon Vertex Detector (FVTX) sensor in which a microscope shows silicon strips positioning and separated at 75 microns [24]. The fundamental silicon detectors that on the principle of doping, typically about 100 micrometres wide contracted silicon bands to fit them into reverse biased diodes. As charged particles enter these bands, they cause slight ionization currents that can be dignified. Accumulating thousands of these detectors all over the place, a collision point in a particle accelerator can produce a specific photograph of what paths particles receipts. Silicon detectors have much more advanced perseverance in tracing charged particles than wire chambers or cloud chambers. Silicon detectors have certain limitations because they require refined freezing to diminish seepage currents (noise source) and are more costly than this grown-up equipment. They also undergo deprivation over each interval of time from radiation.

b. Germanium Detectors

They are typically recycled for gamma spectroscopy in x-ray spectroscopy as well as in nuclear physics. Whereas silicon detectors cannot be denser than a few millimeters, germanium can have an exhausted, profound thickness of centimeters and consequently can be used as an overall concentration detector for gamma rays up to a few MeV. These detectors are also called hyperpure germanium detectors or high-purity germanium detectors (HPGe). Before existing purification techniques were experienced, germanium crystals could not be manufactured with transparency adequate to empower their use as spectroscopy detectors. Contaminations in the crystals trap electrons and holes, tarnishing the performance of the detectors. Consequently, germanium crystals were doped with lithium ions (Ge (Li)), which is imperative to produce an intrinsic region in which the electrons and holes would be accomplished to reach the contacts and produce a signal.

c. Diamond Detectors

Diamond detectors have numerous resemblances with silicon detectors but are anticipated to compromise substantial benefits, in specific a high radiation toughness and precisely low drift currents. At the current time, it is much more expensive and more challenging to assemble.

d. Cadmium Zinc Telluride and Cadmium Telluride Detectors

Cadmium zinc telluride (CZT) and cadmium telluride (CdTe) detectors have been scientifically innovative for gamma and X-ray spectroscopy. The extraordinary firmness of these constituents means they can commendably decrease gamma-rays and X-rays having energy strengths of more than 20 KeV that old-fashioned silicon-beached sensors are incapable of recognizing [25]. The wide-ranging band gap of these resources also means they require extraordinary resistive properties and are proficient in working at room temperature (~295K), corresponding to germanium-based sensors. These detector materials can be used to produce sensors with changed electrode assemblies for imaging and high-resolution spectroscopy. Conversely, CZT detectors are commonly incapable of contesting the resolution of germanium detectors, with more or less of this modification being routinely competent to deprive positive charge-mover transportation to the electrode. Purposes to moderate this consequence have comprehended the development of novel electrodes to reverse the necessity for both polarities of carriers to be self-possessed [26].

e. Automated Detection

Mechanical detection for gamma spectroscopy in typical trials has conventionally been costly since the analysers are a prerequisite to being protected against background radiation. Still, a low-cost auto sampler has recently been accessible for such kinds of investigations [27]. It can be assimilated into diverse mechanisms and instrumentations from numerous manufacturers.

REFERENCES

[1] R.C. Jones, "Erratum: The Ultimate Sensitivity of Radiation Detectors", *J. Opt. Soc. Am.,* vol. 39, no. 5, p. 343, 1949. [http://dx.doi.org/10.1364/JOSA.39.000343]

[2] R.C. Jones, "A new classification system for radiation detectors", *J. Opt. Soc. Am.,* vol. 39, no. 5, pp. 327-343, 1949. [http://dx.doi.org/10.1364/JOSA.39.000327] [PMID: 18131432]

[3] R.C. Jones, "Factors of merit for radiation detectors", *J. Opt. Soc. Am.,* vol. 39, no. 5, pp. 344-356, 1949. [http://dx.doi.org/10.1364/JOSA.39.000344] [PMID: 18144695]

[4] A Handbook of Radioactivity Measurements Procedures (2nd ed.). National Council on Radiation Protection and Measurements (NCRP). 1985. pp. 30–31. ISBN 978-0-913392-71-3.

[5] E. Rutherford, and H. Geiger, "An electrical method of counting the number of α particles from radioactive substances", *Proceedings of the Royal Society (London),* vol. 81, pp. 141-161, 1908.

[6] F. Glenn, *Knoll. Radiation Detection and Measurement.* 3rd ed. John Wiley and sons, 2000.

[7] Geiger Muller Counter. [(accessed on 20 December 2022)]; Available online: https://www.cpp.edu/~pbsiegel/bio431/texnotes/chapter4.pdf?

[8] Nuclear Open-Source Radiation Monitoring System. [(accessed on 20 December 2022)]; Available online: https://www.blackhat.com/docs/us-17/wednesday/us-17-Santamarta-Go-Nuclear-Breaking%20 Radition-Monitoring-Devices-wp.pdf

[9] N. Veall, "A Geiger-Muller counter for measuring the beta-ray activity of liquids, and its application to medical tracer experiments", *Br. J. Radiol.,* vol. 21, no. 247, pp. 347-351, 1948. [http://dx.doi.org/10.1259/0007-1285-21-247-347] [PMID: 18869136]

[10] Korff, S. Das Geiger-Müller-Zählrohr. NTM Zeitschrift Für Geschichte Der Wissenschaften, Technik Und Medizin, 20(4), 271–308, 2012. [http://dx.doi.org/10.1007/s00048-012-0080-y]

[11] Arthur H. Compton – Facts. NobelPrize.org. Nobel Prize Outreach AB 2023. Fri. 6 Jan 2023. Available at: https://www.nobelprize.org/prizes/physics/1927/compton/facts/

[12] The Nobel Prize in Physics 1936. NobelPrize.org. Nobel Prize Outreach AB 2023. Fri. 6 Jan 2023. Available at: https://www.nobelprize.org/prizes/physics/1936/summary/

[13] J.L. Putman, The nuclear handbook: Consulting Editor O.R. Frisch: Newnes, London, 1958. xv + 640 octavo pp. 50s, The International Journal of Applied Radiation and Isotopes, Volume 5, Issue 3, 1959,Page 238, ISSN 0020-708X, Available at: https://www.sciencedirect.com/science/article/pii/0020708X5990238810.1016/0020-708X(59)90238-8

[14] N.N.D. Gupta, and S.K. Ghosh, "A Report on the Wilson Cloud Chamber and Its Applications in Physics", *Rev. Mod. Phys.,* vol. 18, no. 2, pp. 225-290, 1946. [http://dx.doi.org/10.1103/RevModPhys.18.225]

[15] C Chaloner. "The Most Wonderful Experiment in the World: A History of the Cloud Chamber." The British Journal for the History of Science, vol. 30, no. 3, 1997, pp. 357–74. JSTOR, Available at: https://www.jstor.org/stable/4027867 Accessed 6 Jan. 2023.

[16] C.T.R. Wilson, Proceedings of the Royal Society of London, "Series A, Containing Papers of a Mathematical and Physical Character", *Royal Society,* vol. 87, pp. 277-292, 1912.

[17] S.C. Curran, *Counting tubes, theory and applications.* Academic Press: New York, 1949, p. 235.

[18] "Spinthariscope." Merriam-Webster.com Dictionary, Merriam-Webster, Available at: https://www.merriam-webster.com/dictionary/spinthariscope Accessed 6 Jan. 2023.

[19] S. Gao, "Automatic Detection and Monitoring System of Pantograph–Catenary in China's High-Speed Railways", *IEEE Transactions on Instrumentation and Measurement,* vol. 70, pp. 1-12, 2021. [http://dx.doi.org/10.1109/TIM.2020.3022487]

[20] H. Paul, "Real-time automatic interpolation of ambient gamma dose rates from the Dutch radioactivity monitoring network, Computers & Geosciences", *Computers & Geosciences,* vol. 35, no. 8, pp. 1711-1721, 2009. [http://dx.doi.org/10.1016/j.cageo.2008.10.011]

[21] D.A.R. Babu, A. Raman, P. Ashok Kumar, and D.N. Sharma. Portable Micro-R survey meters for environmental gamma radiation monitoring. Radiation Protection and Environment, 26(1-2,pt2), 400-402, 2003.

[22] G.F. Knoll, *Radiation Detection and Measurement.* 3rd ed. Wiley, 1999, p. 365.

[23] Consequences of lake and river ice loss on cultural ecosystem services, Limnology and Oceanography Letters, Lesley B. Knoll, Sapna Sharma, Blaize A. Denfeld, Giovanna Flaim, Yukari Hori, John J. Magnuson, Dietmar Straile, Gesa A. Weyhenmeyer First published: 07 August 2019.[http://dx.doi.org/10.1002/lol2.10116]

[24] M. Carvalho, "Auto-HPGe, an autosampler for gamma-ray spectroscopy using high-purity germanium (HPGe) detectors and heavy shields". HardwareX 4: e00040. 2018. [http://dx.doi.org/10.1016/j.ohx.2018.e00040]

[25] P N. Luke, Unipolar charge sensing with coplanar electrodes -- Application to semiconductor

detectors. United States: N. p., 1995.
[http://dx.doi.org/10.1109/23.467848]

[26] J.S. Kapustinsky, "Production and performance of the silicon sensor and custom readout electronics for the PHENIX FVTX tracker", *Nucl. Instrum. Methods Phys. Res. A,* vol. 617, no. 1-3, pp. 546-548, 2010.
[http://dx.doi.org/10.1016/j.nima.2009.09.036]

[27] M.C. Carvalho, C.J. Sanders, and C. Holloway, "Auto-HPGe, an autosampler for gamma-ray spectroscopy using high-purity germanium (HPGe) detectors and heavy shields", *HardwareX,* vol. 4, p. e00040, 2018.
[http://dx.doi.org/10.1016/j.ohx.2018.e00040]

CHAPTER 5

Particle Accelerators

Abstract: Particle accelerators are devices that increase the speed of elementary particles like electrons or protons. Such particles are accelerated to enormously elevated energies and yield an energetic stream of beams recycled for appreciative identifications about the forces that act upon them (elemental and nuclear behaviour) and fundamental structure wedges of natural surroundings. The present chapter is dedicated to understandings of Linear accelerators along with construction working, calculation of the length of drift tubes, and calculation of particle energy. It also focuses on the cyclotron, its principle, construction, working and applications.

Keywords: Beams, Cyclotron, Linear Accelerator, Particle Physics.

1. INTRODUCTION

On an elementary ground, particle accelerators harvest shaft of rays (beam) using exciting elements that can be recycled for multiple investigations and scientific determinations. It gives an idea about accepting the configuration and dynamic forces acting on the materials and their belongings (chemistry, natural science, nuclear physics, medicinal applications). Stream of accelerated particles can be used to yield beams of secondary particles such as:

- Photons (gamma-rays, x-rays, observable light) are generated from beams of electrons.
- Neutrons are produced beginning from beams of protons by using the spallation neutron source.
- Such kind of subordinate particle streams of beams are used to investigate the materials and their properties in the form of probes [1].

Massive accelerators are used for elementary examination in particle physics. The leading accelerator is LHC, operated by the CERN, the Large Hadron Collider situated in Switzerland, nearby Geneva. It is an accelerator that works on the principle of particle collision, which can speed up the dual-stream of protons beam to an energy of 6.5 TeV and allow them to strike directly, generating centre

Ritesh Kohale, Sanjay J. Dhoble & Vibha Chopra

of mass energies of 13 TeV. Accelerators are also used as synchrotron light sources for learning reduced masses and condensed matter behaviour in physics. Compact particle accelerators are used in multiple applications, together with particle analysis. At the moment, around 30,000 accelerators are under usage and advancement all over the globe [2]. There are dual elementary divisions of accelerators: electrostatic and electrodynamic accelerators [3]. Electrostatic accelerators use static electric fields to speed up particles. The most communal categories are the Van de Graaff and Cockcroft–Walton generators. A moderate pattern of this class that gives a precise determination of the motion of a particle with pictorial representation is the CRT (cathode ray tube). The possible kinematic energy for elementary entities in these approaches is resolute by increasing the accelerating voltage, which is restricted and pulled down by electrical intermission. Additionally, electrodynamic accelerators use varying electromagnetic fields (*viz.*, oscillating radio frequency fields) to speed up and fast-track the particles. Such kinds of particles can penetrate through the corresponding accelerating field with the highest frequency; the production energy is not restricted by the influence of the field with an increase in velocity. This kind of accelerator was primarily recognized in the twentieth century near 1920 and is the base for the most all-encompassing accelerators.

Gustav Ising, Max Steenbeck and Ernest Lawrence are well-thought-out inventors of this arena, recognizing and constructing the initial and practically working LINAC (linear particle accelerator) [4], the cyclotron and the betatron.

1.1. Significant Categories of Particle Accelerator

A widespread selection of particle accelerators is in use these days. The categories of technologies and accelerators are remarkably identified by the speed of particles that are augmented and sped up by the mass of particles conveyed. It has been observed that accelerators for electrons commonly 'appeared' not equivalent to accelerators for bulky or heavy ions and the positively charged protons. The three basic types of particle accelerator are,

1. *Linear accelerator* - Particle acquires energy while moving in a straight line.

2. *Cyclic accelerator* - Particle acquires energy while moving in a circular path.

3. *Static accelerator* - Particle acquires energy while moving in a static path.

1.2. In What Manner Does a Particle Accelerator Performs

Particle accelerators consume electrically powered fields to expedite (accelerate) and intensify the energy of particle beams, which remain strongly directed and focused by magnetic fields. The source makes the elementary particles, such as negatively charged electrons or positively charged protons that are to be speeded. The stream of particles (beam) moves within the evacuated compartment in the metallic beam cylinder. Such evacuation is critical to upholding a midair and any environment without contamination for the particles to move on freely. The directions and focusing of the beam of particles are maintained by precisely maintained electromagnets while it moves through the evacuated cylinder. Electric fields space out all over the place around the accelerator that switches from positive to negative potential at a specified rate of recurrence by maintaining the radio frequency that speeds up the particles in groups. Particles can be focused at an immovable objective, such as a minute unit of metal foil, or dual beams of particles can be struck. Particle detectors disclose as well as record the particles. Moreover, it can also give significant information about radiation formed by the collision between a particle beam and the target.

2. LINEAR ACCELERATOR (LINAC)

(LINAC) A linear particle accelerator is a category of particle accelerator that significantly raises the kinematic energy of subatomic stimulating particles or ions by exposing the charged particles to a succession of oscillating electric potentials laterally along a linear beamline; this technique of speeding up for elementary particles was developed by Leo Szilard.

In linear accelerators, where the fields with increased speed are equivalent to the speed of the particle, the accelerating arrangement is deliberated such that the periodic (phase) velocity of the wave is identical to the velocity of the elementary particles that are to be accelerated.

In 1924 Gustav Ising projected the significant ideologies for such apparatuses [5], while the initial instrument that worked was assembled by Rolf Wideroe in 1928 [6] at the University of Aachen. LINACs ensure numerous solicitations: they produce high-energy electrons and X-rays for medicinal determinations, work as particle incubators for higher-energy accelerators, and are cast off cooperatively to attain the utmost kinematic energy for the elementary particles like positrons and electrons with some other light particles [7].

The line of outbreak and focusing of a LINAC can be influenced by the category of particle pertaining to their nature, the charge, the mass and the spin that is not similar to electrons, protons or any other fundamental ions. LINACs ranges in magnitude, shape and size. One can believe the practical applications from a cathode ray tube (which is a type of LINAC) to the 3.2-kilometre-long (2.0 mi) LINAC [8].

2.1. Construction of Linear Accelerator

1. Fig. (**5.1**) shows the schematics of a linear accelerator. The knowledge of its construction can be resolute by the type of particle that is actually speeded: *i.e.* Protons, Electrons, or Ions.

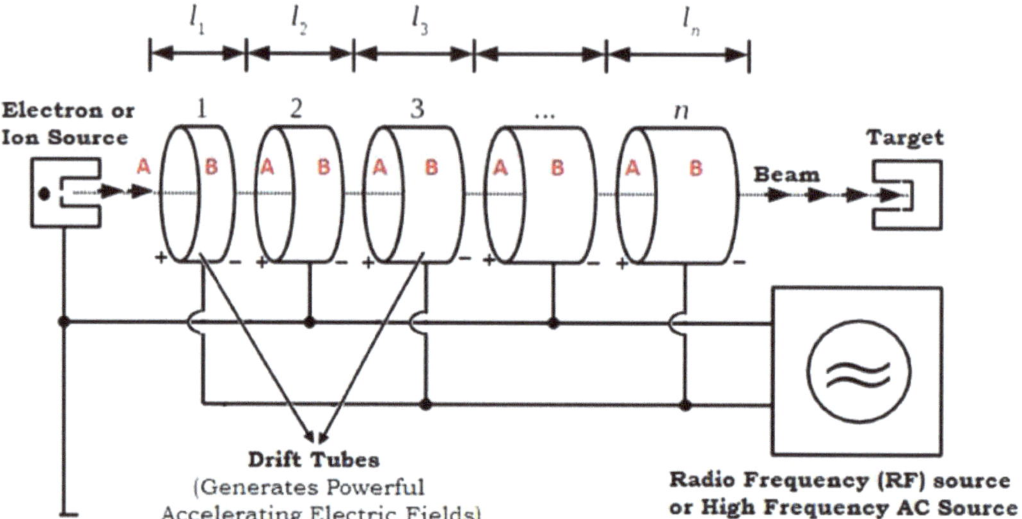

Fig. (5.1). Schematics of a linear accelerator.

2. It consists of the ion source with high voltage, drift tubes of increasing lengths and radio frequency (RF) AC source.

3. Negatively charged Electrons are produced by thermionic emission from a cold cathode, the hot cathode and photon-operated cathode, or specific ion sources that work on the radio frequency.

4. In the interior of the hollow chamber, electrically insulated cylinder-shaped pipes (called drift tubes or an electrode) are positioned, whose distance and linear dimensions change with the distance together with the cylindrical pipes.

5. The linear size of the individual tube is resolute by the rate of recurrence of a particle in the tube and the influence of the exciting source, along with the nature of the particle to be fast-tracked.

6. The drift tubes are smaller nearby the source and longer nearby the target.

7. As the particles passes from horizontal path inside the tube and remain natural and unaffected inside the drift tube, while the frequency of the driving signal and

the spacing between the electrodes are decided so that the maximum voltage appears as the particle moves between the gap.

8. This run-through speed ups the particle, imparting an energy to it with increased initial velocity and, ultimately, kinetic energy.

9. The distance between drift tubes increases along the length of the tube so that the particle stays in phase with the voltage.

2.2. Working of a Linear Accelerator

1. A linear accelerator basically operates based on electrical attraction and repulsion due to applied alternating fields by radio frequency oscillator.

2. Ion source provides essential stream of electrons (beam) which is then fast-tracked in the direction of initial drift tube since they are negatively charged and drift tubes are kept appropriate positive potential.

3. When electrons comes out of tube, in that moment radio frequency (RF) source changes its field to opposite pole. At this moment the first drift tube turn out to be negative and second drift tube becomes positive.

4. When such drifted particles moves towards the succeeding tube, its polarity changes that it has identical polarity as the particle. A third tube, objectively kept afar the second one, becomes charged with the reverse polarity, and the identical mechanism take place. This remains same and continuously go on during the entire process with respect to subsequent tubes.

5. Within a cylindrical metallic conductor, the electric field is null consequently the charge of the tube does not have polar on impact the particle. But as soon as particle appears in the space sandwiched amongst the tubes, it responds to an electric field which energies and push it onward.

6. Negatively charged electrons comes out of tube because of their static mass and ultimately the inertia, hence they are pushed with first drift tube and drifted by the second tube in the similar direction.

7. During this process of successive acceleration tube by tube, the velocity of particle becomes superior and they encounter the longer distance in the equivalent time period. That is the reason why drift tubes must be at extended distance as electrons with extremely high velocity arises nearer to objective or target.

8. The distance between tubes increases along the length of the tube so that the particle stays in phase with the voltage.

9. Noticeably, the dimension and longitudinal elongation (length) of the tubes and the accelerating energy of particle needs to be calculated accurately.

2.3. Calculation of Length of Drift Tube

Consider,

'n' be the total no of drift tubes (*i.e.* n = 1, 2, 3, 4...)

'Ln' .is the length of nth drift tube

'v_{an}' is the velocity of ion (particle) entering or leaving the nth tube

'f' be the frequency of radio frequency oscillator (RFO)

'V' is potential difference applied between adjacent tubes

'm' is the mass of ion (particle)

'q' be the charge of ion (particle)

As we know, the time required for particle to travel within each drift tube is,

$$t = \frac{L_n}{v_n} \quad \textbf{(1)}$$

Time (t) required for the particle to travel through any tube must be equal to half the time period (T) of an applied R. F. voltage. With this condition the particle gets accelerated through each gap.

$$\therefore t = \frac{T}{2} = \frac{1}{2f} \quad \textbf{(2)}$$

Where, f = frequency of R. F. voltage

Equating eq. (1) and (2), we get:

$$\therefore \frac{L_n}{v_n} = \frac{1}{2f}$$

Therefore,

$$L_n = \frac{v_n}{2f}$$

2.4. Calculation of Energy of the Particle

When, 'q' is charge of positive ion and 'V' is potential of RF oscillator

The energy acquired the positive ion in passing one gap = **qV**

Total K.E. acquired by the positive ion after passing through nth drift tube = **nqV**

If, m= mass of positive ion and

v_n = velocity of positive ion when it leaves the nth drift tube

The K.E. acquired by the positive ion $= \frac{1}{2} m v_n^2$

$$\text{i.e. } nqV = \frac{1}{2} m v_n^2$$

$$\therefore v_n^2 = \frac{2nqV}{m} \tag{4}$$

$$\therefore v_n = \sqrt{\frac{2nqV}{m}}$$

Substituting value of v_n in in equation (3) we get,

$$L_n = \frac{v_n}{2f} = \frac{1}{2f}\sqrt{\frac{2nqV}{m}} \tag{5}$$

This is an expression for length of drift tube.

And equation (5) reveals,

$$\mathbf{L_n \propto \sqrt{n}}$$

Since f, q, V and m are constants.

Therefore, the lengths of drift tubes is proportional to $\sqrt{n}$

i.e. L_1: L_2: L_3:.............. :: 1: $\sqrt{2}$: $\sqrt{3}$:.....

Thus the length of drift tubes must be in increasing order so that particle takes constant time to travel through each tube.

3. CYCLOTRON

A cyclotron is a device used to accelerate charged particles to high energies. The first cyclotron was built by Ernest Orlando Lawrence and his graduate student, M. Stanley Livingston, at the University of California, Berkley, in the early 1930's.

3.1. Principle, Construction and Working of Cyclotron

Principle

The fundamental principle on which most of the existing Cyclotrons works is that, when a particle with identical charge moving perpendicular to an applied magnetic field it is pushed in in a trajectory due to magnetic Lorentz force and ultimately particle travels in a circular track.

Assuming above principle, the Cyclotron is constructed on the standard that a positively charged ion can attain adequately enormous energy with a fairly small interchanging potential difference by constructing them to pass away from the identical electric field and again by making use of a resilient external applied magnetic field.

Construction

- It involves a void metallic cylinder separated into dual sections D1 and D2 entitled Dees, sealed off in a vacuum cavity (Fig. **5.2**).
- The Dees D1 and D2 are set aside such that they do not have any contacts and an ion source is positioned at the center in the gap sandwiched between the Dees.
- They are retained in the middle of the pole pieces of a sturdy electromagnet.
- In individual Dee the external applied magnetic field acts at right angles to the plane of the Dees.
- The Dees are linked to a high energy radio frequency oscillator (RFO). In the space splitting the metallic Dees there is a constant electrical field directing from one Dee to the other.
- The moment a charge is released and reposed in the gap it is speeded by the electrical field and transported into one of the Dees.
- The externally applied magnetic field in the Dee sources the charge to track a semicircular path that relocates it back to the gap.

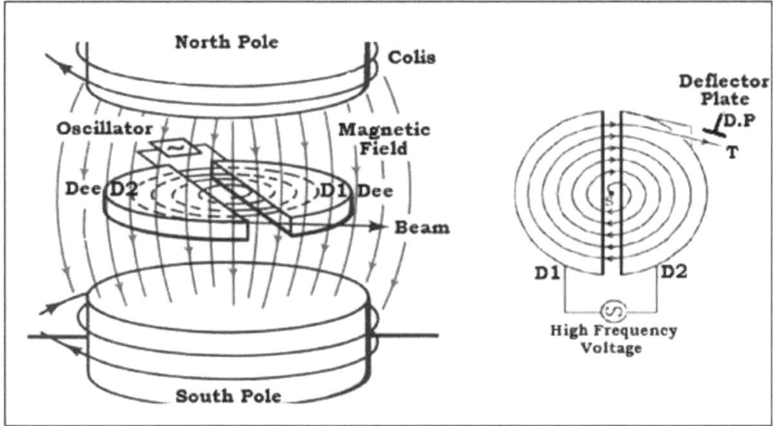

Fig. (5.2). Construction of Cyclotron and accelerated (spiral) path of particle.

Working

- When a positively charged ion with charge q and mass m is released from the ion source, it is speeded up in the direction of the Dee, ensuring a negative potential at that instant.
- Owing to the typical magnetic field, the ion reacts to magnetic Lorentz force and follow a trajectory in a semi-circular path.
- During this time, the resultant ion reaches the gap sandwiched between the Dees, and the polarity of the Dees gets reversed.
- Henceforth the particle is once more speeded and travels into the former Dee with a larger velocity along a circle of a bigger extent in radius.
- Consequently, the particle experiences the movements in a helix or coiled track of increasing radius, and when it arises nearby the edge, it is taken out with the help of a deflector plate (D.P).
- The ions or particles with extraordinary energy are allowable at this time to hit the target T.

When the particle exchanges along a circle of radius 'r' with a velocity 'v', the magnetic (B) Lorentz force gives rise to the essential centripetal force.

Force experienced by the positive ion inside the Dees,

$$F = B\,q\,v$$

As the particle of mass, 'm' moves along a circle of radius 'r' with a velocity 'v' in the Dee, Centripetal force acting on positive ion can be given by,

$$F = \frac{mv^2}{r}$$

Equating both the forces, we get:

$$\frac{mv^2}{r} = Bqv$$

$$\therefore \frac{v}{r} = \frac{Bq}{m} = \text{constant} \quad \textbf{(1)}$$

$$\therefore r = \frac{mv}{Bq}$$

The time taken by particle to describe the semi-circular path can be given by,

$$t = \frac{\text{Distance travelled in a Dee}}{\text{Velocity of particle}} = \frac{2\pi r/2}{v} \quad \textbf{(2)}$$

Exchanging equation (1) in (2),

$$t = \frac{\pi mv}{Bqv} = \frac{\pi m}{Bq}$$

$$t = \frac{\pi m}{Bq} \quad \textbf{(3)}$$

It is clearly anticipated from equation (3) that the time occupied by the ion to define a semi-circle does not depend on:

(i) The radius (r) of the path trailed and (ii) the kinetic velocity (v) of the particle

Hence, the time period of revolution of a particle can be given by,

$$T = 2t$$

$$\therefore T = \frac{2\pi m}{Bq} = \text{constant} \quad \textbf{(4)}$$

So, in a uniform magnetic field, the ion traverses all the circles at exactly the same time.

The frequency of Cyclotron (or rotation of the particle) is,

$$f = \frac{1}{T} = \frac{Bq}{2m\pi} \tag{5}$$

If the high-frequency oscillator is attuned to generate oscillations of frequency as indicated in equation (5), significant resonance takes place.

Such kind of Cyclotron is used to fast-track protons, deuterons and α - particles.

3.2. Cyclotron Radiation

It is electromagnetic radiation discharged by speeding up charged particles driven back by an external applied magnetic field [9]. The magnetic Lorentz force acting on the particles is mutually at right angles to both the magnetic flux and the particles' motion through them, generating an acceleration of particles that origins them to release energy in the form of identical radiations. And as a consequence of the acceleration, they experience as they wind about the outlines of the magnetic field and flux.

The forename of this radiation originates from the cyclotron, a variety of particle accelerators cast-off in the meantime and in the 1930s to produce extremely energetic particles for modification. The cyclotron makes use of the rounded trajectories that stimulating elementary particles reveal in an identical magnetic field. Additionally, the time period of the revolution is unrestricted to the energy of the particles, which is the most favourable condition for the cyclotron to work at a normal frequency [10]. Radiations generated by Cyclotron are a typical combination of entire charged particles passing from magnetic fields and their respective motions in cyclotrons. Plasma radiation from Cyclotron in the astronomical surrounding or nearby black holes and other cosmological occurrences is a significant foundation of evidence around reserved magnetic fields [11].

The cyclotron has progressed in numerous fields depending on different necessities [12].

a. Classic Cyclotron

The preliminary category of the cyclotron, pronounced in earlier sections, which had an identical magnetic field and continuous frequency, is ordinarily old-fashioned. It was insufficient for perfectly nonrelativistic energies in which the particle's rest energy is less than output energy and had focusing complications.

b. Synchrocyclotron

An antiquated mechanism that expanded energies into the relativistic system by weakening the radio frequency of the oscillator as the trajectory of the particle got bigger concerns to retain it in synchronization with the particle. This remained the supreme imposing accelerator in the twentieth century. Before the 1950s most synchrotrons operated in the form of pulses as an alternative for an endless generation of energy, so their beam current (luminosity) was very trimmed.

c. Isochronous Cyclotron (Isocyclotron)

An assembly that encompasses utmost contemporary mechanisms expands productivity and energy into the relativistic system by means of designed portions of the pole to generate an inhomogeneous magnetic field that is strong enough in neighbouring segments to upsurge centripetal force acting directly on the charged particle as it increases relative mass, with substituting-gradient linking to retain the beam precisely focused.

d. Separated Sector Cyclotron

A mechanism in which the magnet is in distinct segments, divided by cavities without field

e. H− Cyclotron

A cyclotron that rapidly accelerates up negative hydrogen ions, which makes it easy to repel the beam out of the instrument. At the beam departure point at the boundary of the dees, a metal foil bands the electrons from the hydrogen ions, transforming them into positively charged H^+ ions. These are determined in the reverse direction by the magnet, so the beam leaves the device.

3.3. Applications of the Cyclotron

For more than a few years, the constructed cyclotrons so far were the paramount foundation of high-energy beams for power purposes as well as nuclear-physics experimentations; a number of cyclotrons are stagnant and are in use for this type of investigation.

- There are essentially two uses for the cyclotron. Firstly it is used for acceleration of charged particles, and on the other hand, it can be modified to speed up any particle with an identical charge.
- Most frequently, it is utilized to accelerate the particles with a positive charge. The most popular selection from entire elementary particles is protons.

Cyclotron is most widely utilized in physics laboratories and in the medicinal field.

- In the medical extent, the cyclotron is emerging as a proton action source.
- Additional medicinal services are being established nowadays to irradiate tissue with the cyclotron by means of high-energy accelerated protons.
- The positively charged proton, identical to gamma rays, has a high perception power that can be excellently adjusted (by 'modifying' the cyclotron) to avoid the destruction of other useful tissues.
- Cyclotrons rays can be cast off for most cancer treatments.
- In plenty of proton therapies, the ion beams originating from cyclotrons can be effectively used. As it can penetrate the body and kill tumors with minimal damage to healthy tissue along their path.
- Cyclotron beams can be used to attack other atoms to yield short-term isotopes that can emit positrons, and hence they are appropriate for imaging such as Positron Emission Tomography (PET).
- The cyclotron is also recycled to generate radioactive constituents that are used as radiation bases that can be embedded.
- Most of the cyclotrons can be cast off in other particle accelerators to get sophisticated energies and a "bigger bang" in the surrounding of collisions.

3.4. Advantages of the Cyclotron

- Cyclotrons are electrically driven particle accelerator that avoids both the excess power and ultimately money, ever since extra expenditure may be distributed to cumulate the effectiveness.
- Cyclotrons harvest highly energetic uninterrupted particle beams at the objective, so it gives a consistent output of high energy.
- The trimness of the cyclotron decreases other outlays, such as its radiation protectiveness, foundations, and neighboring physical structures.

3.5. Limitations of Cyclotron

- Preserving an identical magnetic field over an enormous region of the semicircular metallic discs (Dees) is problematic.
- At extraordinary speeds, comparative deviation of particle mass setbacks the resonance state.
- At elevated frequencies, relative deviance in the mass of the electron is huge, and hence they cannot be augmented by most of the cyclotrons.

REFERENCES

[1] M.S. Livingston, and J. Blewett, *Particle Accelerators*. McGraw-Hill: New York, 1969. [http://dx.doi.org/10.4159/harvard.9780674424340]

[2] Sarah Witman, "Ten things you might not know about particle accelerators", *Symmetry Magazine*.

Fermi National Accelerator Laboratory.

[3] S. Humphries, *Principles of Charged Particle Acceleration.* Wiley-Interscience, 1986, p. 4.

[4] P. Waloschek, Ed., *The Infancy of Particle Accelerators: Life and Work of Rolf Wideröe.* Vieweg, 1994.
[http://dx.doi.org/10.1007/978-3-663-05244-9]

[5] G. Ising, "Prinzip einer Methode zur Herstellung von Kanalstrahlen hoher Voltzahl", In: *Arkiv for Matematik, Astronomi och Fysik.*, 1924.

[6] R. Widerøe, *Über Ein Neues Prinzip Zur Herstellung Hoher Spannungen,* 1928.
[http://dx.doi.org/10.1007/BF01656341]

[7] Celata, C M, Heavy ions offer a new approach to fusion. CERN Cour. 42(6), 2002, 23. https://cds.cern.ch/record/592979

[8] N.C. Christofilos, R.E. Hester, W.A.S. Lamb, D.D. Reagan, W.A. Sherwood, and R.E. Wright, "High Current Linear Induction Accelerator for Electrons", *Rev. Sci. Instrum.,* vol. 35, no. 7, pp. 886-890, 1964.
[http://dx.doi.org/10.1063/1.1746846]

[9] Benjamin Monreal, "Single-electron cyclotron radiation", *Phys. Today,* vol. 69, no. 1, 2016.
[http://dx.doi.org/10.1063/PT.3.3060]

[10] V. A. Dogiel, "Gamma-ray astronomy", *Contemporary Physics,* vol. 33, no. 2, pp. 91-109, 1992.
[http://dx.doi.org/10.1080/00107519208219534]

[11] V. V. Zheleznyakov, "Space plasma under extreme conditions", *Radiophysics and Quantum Electronics,* 1997.
[http://dx.doi.org/10.1007/BF02677820]

[12] A. Chao, *Handbook of Accelerator Physics and Engineering.* World Scientific, 1999, pp. 13-15.
[http://dx.doi.org/10.1142/3818]

CHAPTER 6

Nuclear Reactors

Abstract: The present chapter is thoroughly dedicated to nuclear power reactors to understand its fundamental conceptions. It also envelops the ancient background and constructed nuclear fission reactors, important components of nuclear reactors, thermal reactors, heavy water, moderator reactors, graphite-gas moderator reactors, and accelerator-driven subcritical reactors (ADSR). The Oklo ancient nuclear reactor is the foundation of natural fission reactors.

Keywords: ADSR, Fission, Moderator, Nuclear Reactors, Thermal Reactors.

1. INTRODUCTION

A nuclear reactor uses a controlled chain reaction, *i.e.*, nuclear fission, to generate power. The energy eliminated away in nuclear reactors may be used for the constructive cause and the production of electricity. In 1939, during their experiments on fission reactions, scientists Hahn and Strassman came across the rare-earth elements in uranium after irradiating it with neutrons. O. Frisch and L. Meitner afterward acknowledged this advent as being due to neutron-induced fission of uranium. This finding was scrutinized, monitored and modified with a few more suggestions. Considering that, on December 2, 1942, Enrico Fermi aimed a chain reaction in a device that consisted of a periodic chimney of herbal uranium separated by graphite moderators. Therefore, Fermi confirmed experimentally the idea of the unique length of the chimney to confirm a sequence reaction. This was performed with a very insignificant collective power of the device, ~ 1W. Contemporary electricity reactors accomplish powers of ~ 3 GW. The rise in electricity in fission reactions was in distinction with managed fusion structures.

Steam generated in nuclear reactors finds application in commercial heating or industrial purposes. A few reactors find their application in the production of isotopes for industrial and scientific usage or weapons-grade plutonium. As per

Ritesh Kohale, Sanjay J. Dhoble & Vibha Chopra

the early 2019 records of IAEA [1], there are 226 nuclear studies reactors and 454 nuclear power reactors being used worldwide [2, 3].

1.1. Principle of Operation

The fundamental Nuclear reactors exchange the energy generated by controlled nuclear fission into thermal energy for further conversion to electrical or mechanical systems similar to thermal power stations that produce electricity from burning fossil fuels.

1.2. Fission

Nuclear fission proceeds when a huge fissile nucleus with uranium (235U) or plutonium (239Pu) captivates a neutron. A bulky nucleus splits into two or extra lighter nuclei during a fission reaction emitting gamma radiation, kinetic energy and free neutrons. The ratio of these neutrons may get consumed by other fissile atoms, which results in the acceleration of additional fission reactions, due to which additional neutrons are released, termed as a nuclear chain reaction. The nucleus of a uranium (^{235}U) atom absorbs a neutron and liberates free neutrons and fast-transferring lighter elements called fission products. However, reactors and nuclear weapons both proceed through a nuclear chain reaction; the rate of reactions in a reactor is more than in a bomb. An illustration of an induced nuclear fission occurrence is shown in Fig. **(6.1)**.

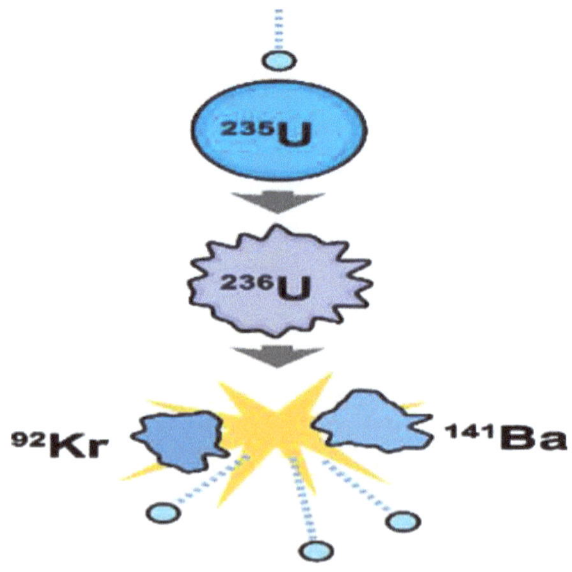

Fig. (6.1). An illustration of an induced nuclear fission event. Courtesy: "Wikimedia Commons." Originally created by Fast fission in Illustrator. Source: https://commons.wikimedia.org/w/index.php?curid=486924.

To regulate such a nuclear chain reaction, control rods encompassing neutron moderators and neutron contagions can modify the percentage of neutrons that will cause additional fission [4]. If monitoring or instrumentation suffers from unsafe conditions, nuclear reactors normally have spontaneous and physical systems to close the fission reaction [5].

1.3. Generation of Heat

There are several techniques to generate heat in the reactor core: Thermal energy is transformed from the kinetic energy of fission products when nearby atoms strike these nuclei. Some of the gamma rays produced are absorbed by the reactor during fission and transformed this energy into heat. The formation of heat occurs by the radioactive decay of fission products and materials activated by neutron absorption. This decay source will subsequently continue for some time, even if the reactor is shut down [6, 7].

1.4. Fission Reactions Rate Mechanism

The fission reaction rate within a reactor center can be adjusted by regulating the quantity of neutrons capable of setting off additional fission movements. Nuclear reactors have diverse strategies to regulate the neutron flow and hence can regulate the reactor's power output. The control rods aid in regulating the rate of the neutron-induced fission reaction. Control rods act as neutron poisons and consequently engross neutrons. When a control rod is inserted deeper into the reactor, it absorbs extra neutrons than the material it transfers (the moderator). Due to this, few neutrons are accessible to cause further fission reactions. On the other hand, if the control rods are completely removed or pushed upwards, the rate of fission reaction will increase, and hence a large amount of power will be generated.

1.5. Cooling System

Most commonly, the coolants in a nuclear reactor are water and sometimes a gas or liquid sodium or lead (liquid metallic) or molten salt. This coolant runs through the reactor core to captivate the heat generated in the reactor. The heat generated from the reactor is used to generate steam.

Most reactor systems use a cooling system alienated from the water that will be boiled to generate pressurized steam to run the turbines. Although, just like traditional boiling water reactors, in nuclear reactors, the water for the steam turbines is boiled openly by the reactor core [8].

1.6. Power Generation

The energy produced in the fission process can be converted into usable energy as it generates a sufficient extent of heat. A practical and popular technique to acquire this thermal energy is to boil water to yield pressurized steam to run a steam turbine that runs an alternator and generates electricity for practical applications [9].

2. IMPORTANT COMPONENTS OF NUCLEAR REACTOR

- **Fuel (Reactor Core):** Fissionable materials like ${}^{235}_{92}U$, ${}^{238}_{92}U$, ${}^{239}_{94}U$ are used in the form of rods about 2 to 3 cms in diameter.
- **Moderator:** Moderators are lighter nuclei that control the speed of fast-moving neutrons by elastic collision. Generally used moderators are water, heavy water and graphite and beryllium oxides.
- **Coolant:** By circulating coolant through the reactor, the heat produced in the nuclear reactor gets removed. The most commonly used coolants are cold water, liquid oxygen, liquid metals, and gases, including CO_2.
- **Control Rods:** A chain reaction is required in a nuclear reactor, which absorbs excess neutrons. These control rods are inserted in the core of the nuclear reactor. This reduces the multiplication factor of neutrons to a very small value. Frequently used material for control rods is Cadmium or Boron.

The schematics of a typical nuclear power reactor are shown in Fig. **(6.2)** below.

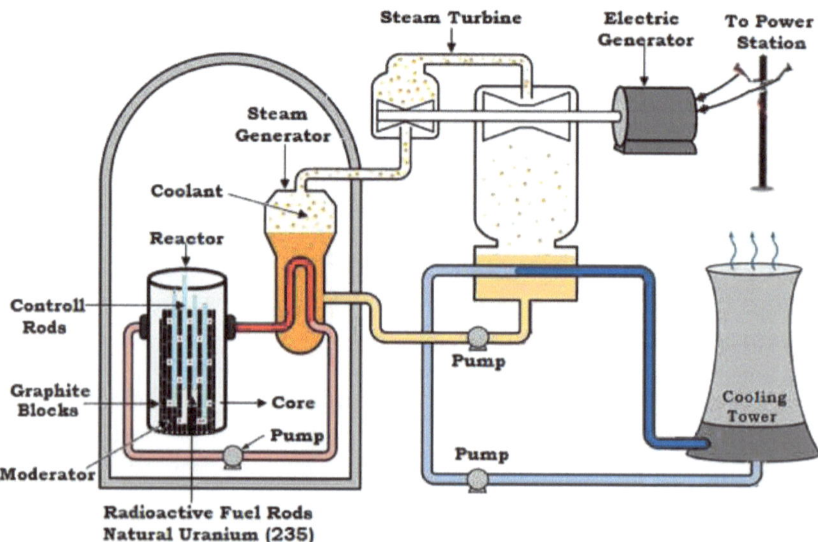

Fig. (6.2). Schematics of Nuclear Power Reactor.

2.1. Construction of Nuclear Reactor

- Nuclear reactors core has uranium (^{235}U) in the form of cylindrical rods. These rods are dipped in a liquid moderator.
- Whenever one neutron strikes this uranium rod, nuclear fission reaction starts, and three fast-moving neutrons are produced.
- The geometry of the core is such that only one out of three neutrons that are emitted strike the next rod making the reaction a controlled one.
- An excess amount of energy is released into the core.
- Water at very high pressure is passed through to extract energy from the core.
- Steam generators produce steam as hot water is passed through them.
- This steam is used for turbines which generate electricity.
- This process will continue until the rods' uranium does not get over. Then the rods have to be replaced in the nuclear reactor.

2.2. Advantages

- The energy released is extensively huge.
- Requires fuel in minute quantity.

2.3. Disadvantages

- Fuel used due to its radioactivity is extremely hazardous to all life forms.
- Collection of radioactive waste.

2.4. Significant Types of Nuclear Reactor

2.4.1. Thermal Reactors

This is the most commonly used category of nuclear reactors. They are known as Pressurized Water Reactors (PWRs). In PWRs, the pressurized water acts as the moderator and the coolant.

The fuel rods are made with ceramic pellets of UO_2, usually amended to around 3% in ^{235}U.The pellets are shielded inside zirconium tubes, lowering the neutron absorption cross-section [10]. The tubes are settled in steel-structure reels inside which the coolant flows. The rolls are protected and are settled perpendicularly in a stable alignment named the core structure, as shown in Fig. (**6.3**). The core is inside an enormous steel pressure vessel.

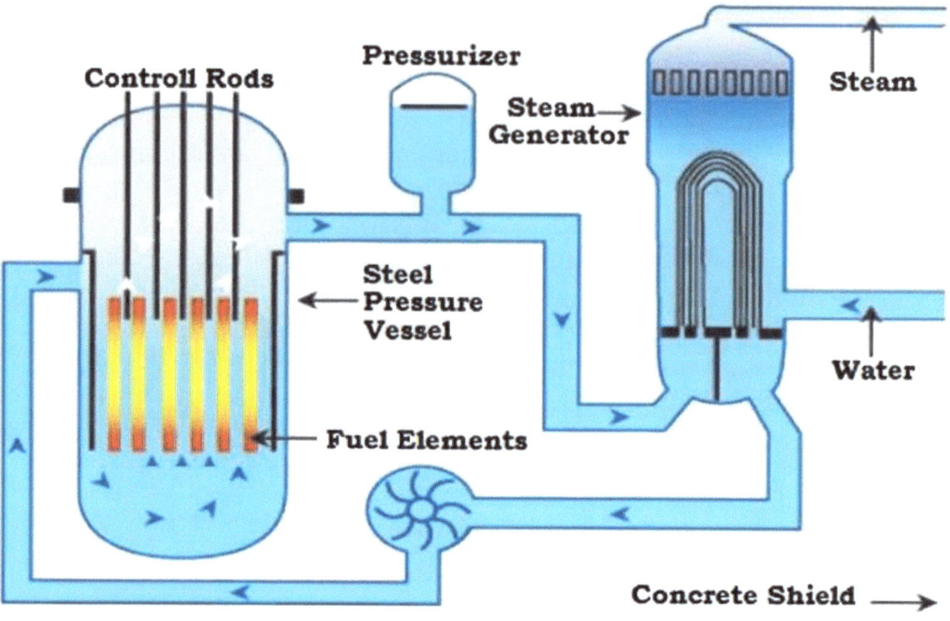

Fig. (6.3). PWR reactor consists of fuel element holder, the control rods filled, the steel structure contains 264 zirconium tubes holding pellets of UO_2.

The concentrated approved power is determined by the prerequisite that the temperature inside fuel rods should be less than the melting point of UO_2, *i.e.* 3100 K. Assuming the neutron fluctuation is persistent inside the tube, the radiated heat per unit volume, *Q*, is self-regulating irrespective of position. The unification of the heat transfer equations gives,

$$Qa^2 = 4k(T_{max} - T_s) \quad \mathbf{(1)}$$

Where 'a' is the radius of the tube, T_{max} is the optimum temperature within the tube, T_s is the surface temperature of the tube, and k is the coefficient of conduction of the uranium oxide.

By using following statistics,

$T_{max} = 1800\ ^0C$,

$T_s = 450\ ^0C$ and

$k = 0.06W/cm/K$,

We have,

$Qa^2 = 300W/cm$.

For a tube diameter of 1 cm, we can have,

$$Q_\rho = 40Wg^{-1}.$$

The density of the oxide is $\sim \rho = 10gcm^{-3}$. The size of the reactor is consequently decided by using the cooling conditions. Hence for 3000 MW, essential the fuel mass required will be:

$$M_{Fuel} = \frac{3000\ MW}{40Wg^{-1}} = 75\ ton$$

This resembles 10^5 m of fuel tubes with a diameter of 1 cm.

In this reactor, water works as both moderator and coolant fluid. Its maximum temperature in the surrounding area of the zirconium alloy tubes is 300 ^{0}C. To stop the water from boiling, it is provided with constant pressure, *i.e.*, up to 150 atmospheres. It flows and provides a heat exchanger, which leads to a subsidiary water circuit that produces steam to drive turbines for electricity generation.

The maximum yield with which the thermal energy of the steam can be converted to work on the turbines and then to electrical energy is controlled by Carnot's principle,

$$\epsilon = 1 - \frac{T_1}{T_2},$$

Where T_1 and T_2 are the final and initial temperatures of water, respectively. In general, temperature T_1 is fixed by a cold source of water such as the sea or river. There are particular alternatives to these reactors depending upon the change in moderator and coolant [11].

2.4.2. Heavy Water Moderator Reactors

Regardless of its high price, heavy water is preferred as a moderator in the system of thermal-neutron reactors. In such reactors, natural uranium is used as a fuel, which is an advantage in the system of thermal-neutron reactors; the pellets are placed in pressurized tubes where heavy water flows at a temperature of 200 ^{0}C (pressure of 90 atm). Heavy water is used as both moderator and coolant.

2.4.3. Graphite-gas Moderator Reactors

In these systems, the natural metallic uranium is used as a fuel. The graphite works as a moderator, and CO_2 is a coolant. Whereas, some drawbacks of the above reactors include using metallic uranium with a low melting point confines the primary temperature and, hence, the thermal efficiency. The slighter moderation power leads to greater sizes than water reactors for the equivalent power.

2.4.4. Fast Neutron Reactors (FNR)

Fast neutron reactors (FNR) are industrialized steps away from conservative power reactors. They compromise the outlook of immensely well-organized consumption of uranium resources and the capability to burn actinides, the long-lasting constituent of sophisticated nuclear byproducts.

2.4.4.1. FNR History

Basically, FNR was considered to burn uranium more effectively and thus outspread the world's uranium assets. From starting, nuclear scientists assumed that present-day reactors were operated basically with ^{235}U broken, less than one percent of the energy accessible from uranium. Premature estimations suggest that uranium resources were uncommon in numerous countries to get on widespread development programs for nuclear reactors. On the other hand, important practical and materials difficulties have come across, and environmental considerations indicated by the 1970s that uranium shortage would not be a distress for a particular time. Due to both issues, in the 1980s, it was clear that FNRs would not be commercially economical with current light water reactors for a specific period.

2.4.4.2. FNR Operation Principal

Using an intermediate productive nucleus such as ^{238}U or ^{232}Th, fast neutrons emit additional nuclear energy, and hence fast neutron reactors can be used as breeders. Two neutrons per fission are produced in fast-neutron (^{239}Pu) reactors in addition to it. (See Table **6.2**). In fast neutron reactors, the most significant fact is to maintain the chain reaction; hence, amongst the two neutrons, one neutron can be recycled to conserve the chain reaction, and the others can produce further ^{239}Pu using neutron absorption on ^{238}U. If the possibility for this to materialize is necessarily nearby to unity, the ^{239}Pu demolished by fission can be substituted by a ^{239}Pu produced by radioactive capture on ^{238}U. The ultimate result is that ^{238}U is the actual fuel of the reactor.

The resulting observations are in order. Think through, for the assertion, a breeder using the productive nucleus ^{238}U. The fissile nucleus is ^{239}Pu for two reasons.

Initially, it is prepared in neutron absorption by ^{238}U, which leads to a closed cycle, in addition, it is produced in large quantities in nuclear expertise, and however one can only depend on the natural resources of ^{235}U.

The probability of limitation must not be too small compared to the possibility that the numerous nuclei in the medium undergo so that productive confinement of a neutron should produce an appreciable amount of the fissile ^{239}Pu inside the fuel. This probability depends both on the physical scheme of the fuel elements and on the quantities of ^{239}Pu, ^{235}U and ^{238}U. It can be estimated in terms of the number of numerous nuclides, fission cross-sections of ^{238}U and its confinement. The identical feature seems for the ^{232}Th-^{233}U pair. This is why fast neutron reactors are used in breeders. Additionally, this also clarifies why in ^{238}U - ^{239}Pu breeders, the design of fuel elements comprises in a central core of ^{239}Pu enclosed by a covering of ^{238}U exhausted in ^{235}U in order to lower the quantity of fission in the peripheral productive region. Consequently, fast neutron reactors are used as breeders, and two fertile fissile pairs are *a priori* possible, $^{238}U -^{239} Pu$ and $^{232}Th -^{233} U$. Contemporary nuclear industry is concerned with the first choice since the thermal reactors create plutonium which is divided in the fuel processing operation. The $^{232}Th -^{233}U$ couple finds application and has many advantages, among which is that it does not lead to considerable quantities of hazardous trans-uranium elements such as americium and curium. These "minor actinides" are produced in consequential amounts and are hazardous, as they have half-lives, and so activities, lying in life-threatening regions, are neither small enough to decay adequately on a human scale nor long enough to be passed over, such as natural uranium and thorium. The half-lives of some of these isotopes are 432 years for 241Am, 7400 years for 243Am and 8500 years for 247 Am. Fig. (**6.4**) reveals the schematics of a hybrid fast neutron reactor. Fig. (**6.5**) indicates the transverse view of fast neutron reactors.

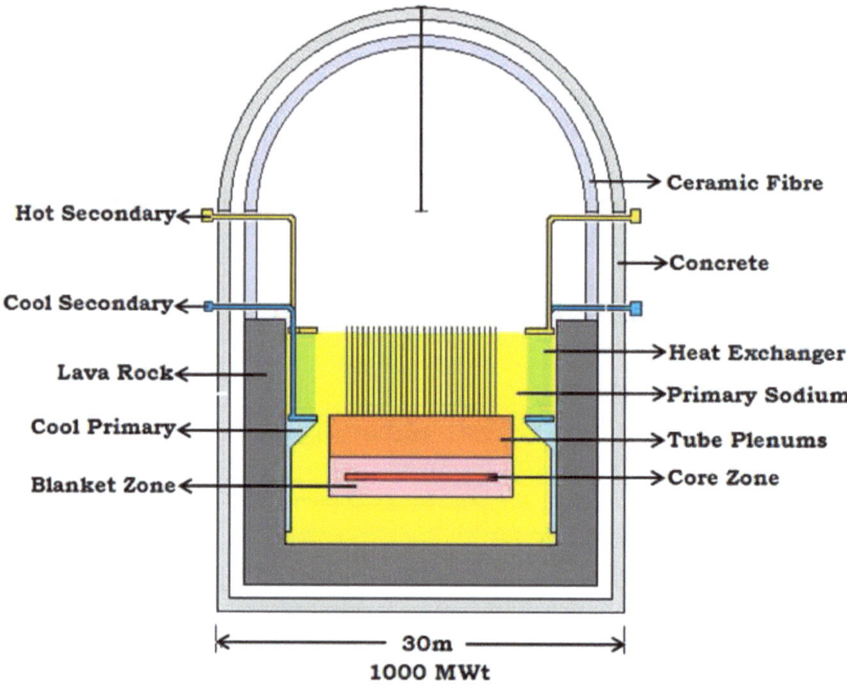

Fig. (6.4). A possible schematics of a hybrid fast neutron reactors.

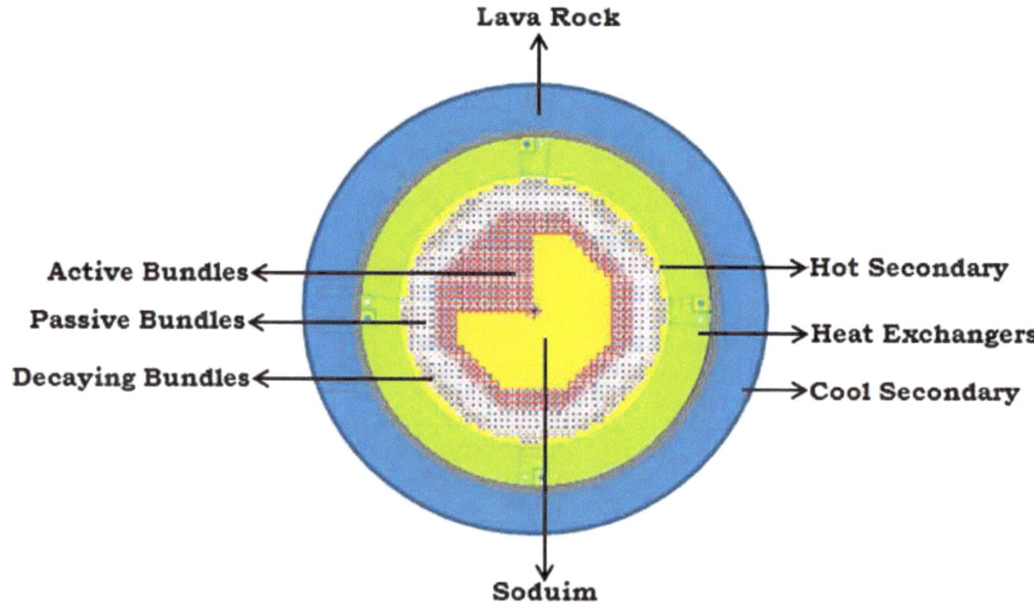

Fig. (6.5). Transverse view of Fast neutron reactors.

2.4.4.3. The Use of Sodium

Coolant fluid with only heavy nuclei is required by a fast neutron reactor to avoid neutron-energy losses. For this liquid, sodium is used, and water is avoided. Molten sodium has identical thermal interchange properties. Molten sodium melts at 98^{0}C, boils at 8820C, and is used at a temperature of 5500C, and there's no need of pressurizing, which is beneficial for the mechanical beginning. Furthermore, it has hydraulic belongings comparable to water at room temperature, which is a valuable concurrence for materials analysis. Another great feature is that sodium can be used at more sophisticated temperatures than water, advancing the thermal/electric alteration factor. At the same time, sodium has a disadvantage in that, due to its highly reactive nature, it burns naturally in air and water. It is a huge drawback given safety precautions. Moreover, due to neutron stimulation, sodium becomes radioactive. Consequently, two circuits are essential. In order to prevent the prime circuit from leaking into the subordinate circuit, the final is maintained at a higher pressure than the initial.

2.5. FNR Features

- Liquid sodium cooled FNRs provide various significant applications compared towater-moderated reactors and security benefits in inhabitant power reactor uses. Such as:
- Additional 100-fold less natural uranium feeding per kWh.
- Neutralizing waste is not produced.
- FNR exhibits low primary sodium pressure;
- FNR achieves higher primary sodium temperatures.
- Inert high-temperature chain reaction ends.
- Ordinary transmission of subordinate heat conveyance fluid adequate for elimination of fission product decay heat.
- FNR allows a high output power gradient rate.
- FNR are suitable for speedy restoration by element replacement.
- It has extensive operational lifetime is exhibited.
- It has exceptional stoppages for refilling.
- FNR could gain extraordinary capability due to many self-governing.
- FNR could have potential applications for making fission fuel ^{233}U from thorium and fusion fuel ^{3}H from ^{6}Li.

2.5.1. Fuel Cycle in FNR

Fast neutron reactors (FNRs) cooled by liquid sodium are predominantly operated by the ample uranium isotope ^{238}U. The chief FNR nuclear process transforms the uranium isotope ^{238}U into the plutonium isotope ^{239}Pu and then fissions the ^{239}Pu. Most of the resultant fission have insignificant half-lives. There exists an additional fission neutron that:

i. If absorbed by an ample ^{238}U atom in the absolute yield a ^{239}Pu atom.
ii. If absorbed by an ample thorium (^{232}Th) atom in the absolute yield a fissionable uranium (^{238}U) atom; or
iii. If absorbed by a lithium isotope (^{6}Li) atom in the absolute yield a mix of tritium (^{3}H), rare helium isotope (^{3}He) and common helium (^{4}He) atoms.

Over a period of one to four decades, the fuel in an FNR is recycled to:

i. Remove fission products.
ii. Replace the removed fission product quantity by an equal quantity of new ^{238}U.
iii. Transfer other transuranium actinides and ^{239}Pu, ^{233}U from the reactor blanket fuel rods to the reactor core fuel rods.
iv. Produce ^{3}H and ^{3}He atoms for consumption in fusion power systems.

The time concerning consecutive fuel recycling stages is recognized as the fuel cycle.

Through each FNR fuel cycle, the Pu fraction in the core fuel reduces from 20% to 12.5% by a weight equivalent to 15% core fuel burn up and 7.5% transformation of Uranium into Plutonium in the core fuel. With ^{238} U absolute fuels, the fuel recycling process produces additional core fuel rod material, which permits a constant increase in the FNR fleet. With ^{232}Th absolute fuel, the fuel reutilizing procedure can endure the current FNR fleet but does not permit a noteworthy increase in the FNR fleet.

2.6. Accelerator-Driven Subcritical Reactor (ADSR)

An accelerator-driven subcritical reactor is a nuclear reactor designed by combining a significantly subcritical nuclear reactor core with a high-energy proton accelerator. It would use thorium as a fuel, which is richer than uranium. The neutrons required for satisfying the fission process would be delivered by a particle accelerator generating neutrons by spallation. These neutrons stimulate the thorium, allowing fission without demanding to make the reactor critical. One advantage of such reactors is the moderately short half-lives of their waste products.

2.6.1. Operational Principal in ADSR

The high energy proton beam influences a molten lead target inside the core, fragmenting or "spallating" neutrons from the lead nuclei. These spallation neutrons convert fertile thorium to fissile uranium-233 and motivate the fission reaction in the uranium. Fig. (**6.6**) indicates the schematics of Accelerator-driven subcritical reactor (ADSR).

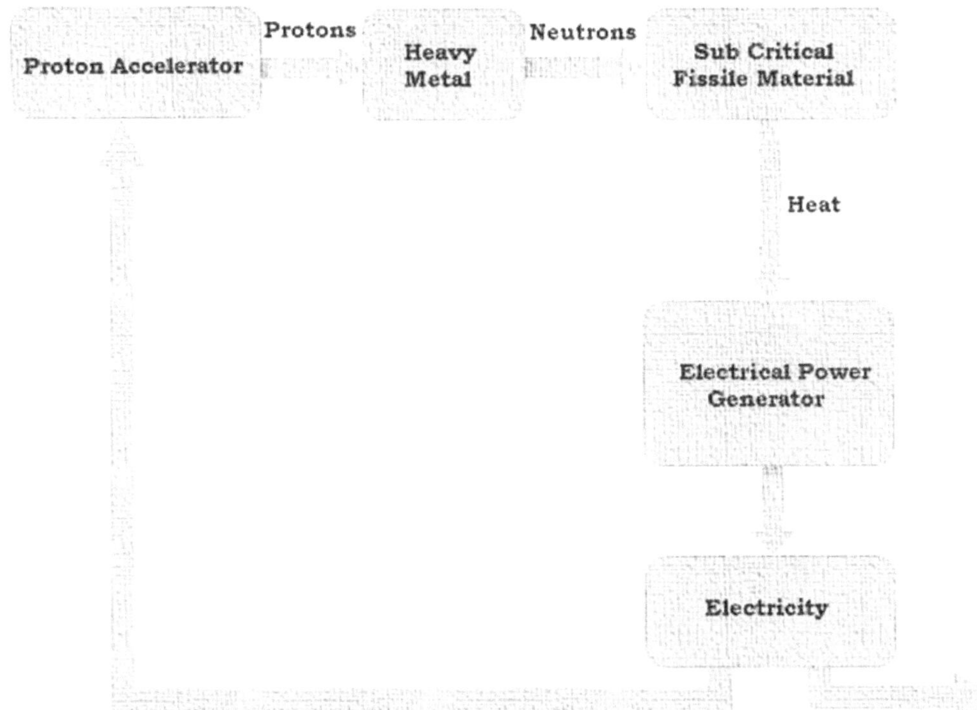

Fig. (6.6). Schematics of ADSR.

A particle accelerator acts as the most appropriate neutron source for such kind of system, which aims at a beam of charged particles on a target. This yields neutrons over (spallation) reactions that break nuclei in the target. The target is enclosed by a sub-critical blanket enclosing fissile material, as shown in Fig. (**6.6**). Such accelerator-coupled detectors are regularly called *hybrid reactors* [12]. Fig. (**6.7**) indicates the scheme of a hybrid reactor.

The proton beam is focused and strikes a region having molten lead which is enclosed by thorium fuel. The heat produced causes the lead to augmentation and rising toward heat exchangers.

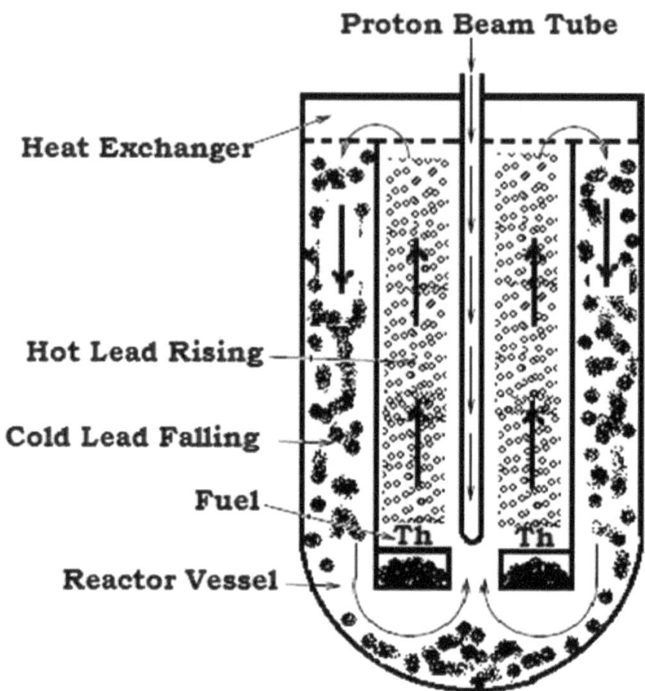

Fig. (6.7). A probable scheme of a hybrid reactor.

2.6.2. The Spallation Target and Sub-critical System

The collaboration of protons with energies E_{proton}> 100MeV in a target adequately dense to break the beam, which results in increased emission of neutrons. They are formed with the primary proton-nucleus reaction and by auxiliary interactions of secondary particles (π, p, n…) within the dense target. This results in a cascade enclosed in the target chamber, the length of the chamber is approximately taken equal to the stopping distance of the incident beam. To obtain an optimum number of neutrons to leave the target, the beam's diameter is adjusted and acts together with the exterior material.

The mean number ν_p of spallation neutrons emitted per incident proton is observed to be,

$$v_p \sim 30 \times E_p$$

where E_p is the incident beam energy in GeV. The energy spectrum of the neutrons released goes from a few keV to the beam energy permitting their creation mechanism (fission, spallation, vaporization). It should be noted that near to the ground energy part (1–2 MeV) is preferred by a dense target, compared to the excited (optimum) energy part related to direct spallation.

2.6.3. The Sub-critical System

The dense target is enclosed by a sub-critical system measured by v and k, the amount of neutrons for each fission before and after adaptation for absorption and geometrical losses. The v_p neutrons added into the sub-critical system by proton are multiplied by k in sequence. The aggregate neutrons per incident proton are equal to:

$$\mathrm{N_t} = \mathrm{v_p}(1 + \mathrm{k} + \mathrm{k}^2 + \cdots) = \frac{\mathrm{v_p}}{(1 - \mathrm{k})}$$

Between these N_t neutrons, $(N_t - v_p)$ are liberated by means of fission. In this deference, we can conclude that the greater the value of $k < 1$, the higher the percentage of fission neutrons in the sub-critical medium. Therefore, the neutron energy spectrum will be calculated further by the sub-critical medium than by the preliminary number of the primary neutrons. Subsequently every single fission yields v neutrons, the aggregate number of fissions for each incident proton is equivalent to,

$$\mathrm{N_{Fission}} = \frac{(\mathrm{N} - \mathrm{v_p})}{\bar{\upsilon}} = \frac{\mathrm{v_p}}{\bar{\upsilon}} \frac{\mathrm{k}}{1 - \mathrm{k}}$$

The expansion of the system, *i.e.* the energy liberated in fission divided by the incident energy of proton can be given by,

$$\mathrm{G} = \frac{\mathrm{N_t E_{Fission}}}{\mathrm{E_p}}$$

$$\mathrm{G} = \frac{\mathrm{E_{Fission}}}{\mathrm{E_p}} \frac{\mathrm{v_p}}{\bar{\upsilon}} \frac{\mathrm{k}}{1 - \mathrm{k}}$$

Where, $E_{Fission} \sim 200MeV$ is the energy liberated in each fission.

In association with above equivalence the total thermal power can be estimated as,

$$\text{Power}(P)_{MW} = E_{Fission}(\text{MeV})\ I(A) \frac{\nu_p}{\bar{\upsilon}} \frac{k}{1-k}$$

Where, I (A) is the beam current. Moreover as criticality (k = 1) is approached the intensity of beam essentially decreases to reach a given power of the reactor.

2.6.4. The High Intensity Beam Accelerator

The intensity of an extraordinary energy beam for a system whose power is comparable to a typical reactor of 3000 MW (thermal) lies between a few mA to 350 mA. High-tech accelerators have slightly a smaller amount of power. The utmost efficient nuclear reactors are either (LINACS) linear accelerators functioning in uninterrupted mode, (*e.g.* 1 mA, 800MeV, Los Alamos, The Linac) or cyclotrons, (*e.g.* 0.8 mA 600 MeV, the PSI cyclotron).

The industrial possibility of attending the essential power needs advancement in numerous domains.

- Utmost care is taken to restrict beam losses to precisely small values (*i.e.* of the order of 10^{-8}/m for a LINAC) in order to avoid galvanizing the structures of the accelerator up to a level where physical involvements turn out to be unbearable thing.
- The efficiency of the RF system (radio frequency) must be essentially amplified in order to acquire the greatest possible efficiency as well.

The scheme of a hybrid reactor shown in Fig. (**6.7**) is established on the basis of thorium cycle using fast neutrons. The fissile material is ^{233}U and the fertile material is ^{232}Th. The scheme is entitled as an energy amplifier since the gain expected is very large at that stage (*i.e.* k = 0.98). By using its regular convection Molten lead finds application and is used both as spallation source and as coolant. The benefits of such a system are as follow:

- The thorium cycle yields little plutonium and heavier elements. For the reason it is considered as “cleaner” cycle than the Uranium cycle.
- Thorium is two times as rich as uranium and does not require isotopic separation.

- Thorium masters the technology of molten lead and plutonium avoids the drawbacks of liquid sodium.
- The accelerator leads to an enhanced mechanism of the system that recognizes the velocity of reactions (of the order of a microsecond) associated with the mechanical usage of control bars (of the order of one second).

2.7. The Oklo Ancient Nuclear Reactor

Oklo Reactor or Oklo Mines, located in Oklo, Gabon, on the west coast of Central Africa, is assumed to be the only natural nuclear fission reactor. Oklo involves sixteen sites at which self-filling nuclear fission reactions are supposed to have taken place around 1.7 billion years ago, and ran for hundreds of thousands of years. It is expected to be more or less under 100 kW of thermal power for the period of that time [13].

2.7.1. Ancient Background

A natural nuclear fission reactor is situated in the Central African state of Gabon (see Fig. **6.9**). As indicated in Fig. (**1**), France Villian is the sector of geological development in which the uranium and ferrous deposits are broken at Oklo. This is a sandy progression enclosed within the intermediate pre-Cambrian sink, which shields an area of about 35000 square kilometers in Gabon. It covers a crystalline foundation along with the interaction with which several uraniferous layers have been found. The uranium in these layers may have been resulting from leakage of the foundation rocks. Dating amounts designates an age of 1740 ± 20 million years. Taking advantage of the Oklo deposit is carried out by a Gabonese community, namely "La Compagnie des Mines d'Uranium de Franceville (COMUF)," with the contribution of the French capital.

Both man made nor assembled from contemporary concrete and steel, Oklo is a nuclear reactor sector of sandstone, granite, and uranium rocks originated in nature spontaneously (see Fig. **6.8**). In May 1972 it was first reported and exposed by a nuclear fuel-processing plant in France. The French Atomic Energy Commission investigated a discrepancy in natural Uranium-235, around 200 kilograms of U-235, which is adequate and equivalent to building 6 nuclear bombs. Moreover, it has been observed that the percentage of U-235 in the sections of uranium from the Oklo deposit in Gabon (0.717%) was somewhat precast from the 0.720% originating in uranium in another place of the earth's layer. The transformation originates from the uranium reaction billions of years ago, which altered the amount of uranium ore in Oklo. Fig. (**6.9**) indicates schematics of geological conditions in Gabon leading to traditional nuclear fission reactors [14].

Fig. (6.8). Geographic location of Oklo Nuclear Reactor on Earth.

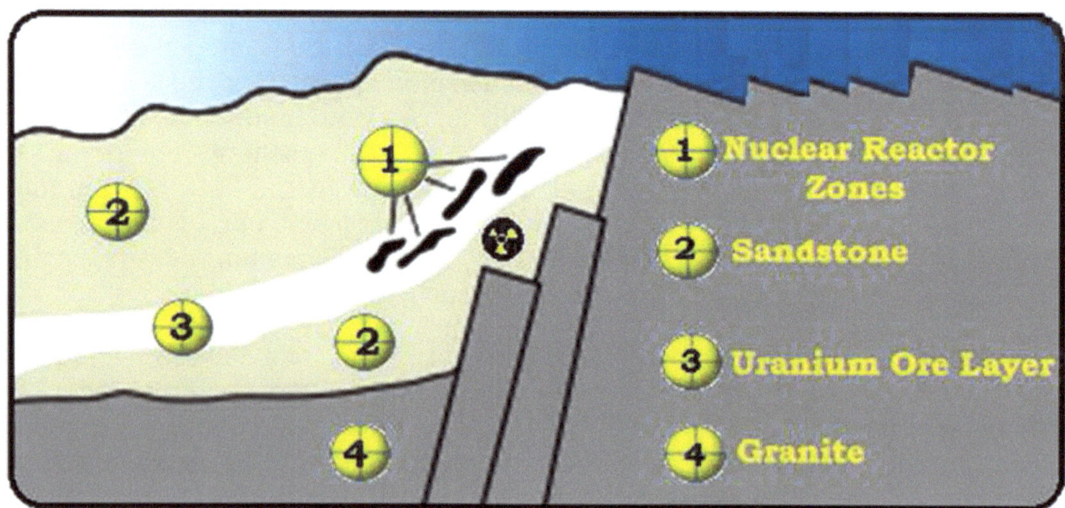

Fig. (6.9). Schematics of geological condition in Gabon leading to traditional nuclear fission reactors.

2.7.2. Foundation of Natural Fission Reactors

In 1953, scientists from the University of Chicago issued expectations that certain amount of uranium deposits may possibly be operating as self-operating nuclear fission reactors, this fundamental prediction was put forward before 19 years the discovery of the Oklo natural nuclear reactor. A significant Scheme by a chemist named Paul K. Kuroda from the University of Arkansas established three circumstances essential for a natural fission reactor.

i. First, the width of the uranium deposit needs to be more than two-thirds of a meter. This permits neutrons to adequately travel to be absorbed and carry on the nuclear chain reaction before leaving the uranium deposit.
ii. Second, the percentage of Uranium-235 must be above 3% in the uranium rocks to fuel a nuclear reactor adequately. It should be noted that U-235 degenerates over time; consequently, two billion years ago, there was sufficient U-235 in Oklo to permit a natural fission reactor to exist. Currently, the composition of U-235 is too weak.
iii. Third, a neutron moderator (for example, water) is required to switch the nuclear reaction - one without considerable lithium, boron, or whatever that will absorb excessive neutrons.

2.7.3. Innovations from Oklo

Oklo provides a time machine to observe how nuclear waste can be controlled. Scientists have analyzed how nuclear waste from reactions affected neighboring regions two billion years ago. Oklo has established that aluminum phosphate minerals are capable of enclosing radioactive gases that are usually more dangerous derivatives of nuclear reactions, such as Xe-135 and Kr-85, for billions of years.

There have also been novel hypotheses regarding the time variation of the fine-structure constant alpha by expending "a Maxwell-Boltzmann low energy neutron spectrum." These are recognized on an investigation from isotopic analysis of samples from Oklo.

The most significant consequence Oklo nuclear reactor is that in the thousands of years of operating as a nuclear reactor, Oklo never had a breakdown or burst [15]. The protection of contemporary nuclear power plants can be cognizant by billions of years in the environment. The mixture of aluminum phosphate particles to set-up radioactive materials and the groundwater to control the reaction permissible for an extremely safe reactor.

2.8. Existing Technologies

Normally reactors are used for examination and training, materials testing, or the creation of radioisotopes for medicine and industry. These are much smaller than power reactors or those thrusting ships, and many are on university campuses. About 280 such reactors are operating in 56 countries. Some operate with high-enriched uranium fuel, and international efforts are underway to substitute low-enriched fuel [16].

2.8.1. Pressurized Water Reactors (PWR)

These reactors use a pressure vessel to contain the nuclear fuel, control rods, moderators, and coolant. The hot radioactive water that leaves the pressure vessel is twisted from side to side with a steam generator, which heats a nonradioactive coil of water to steam that can move turbines. They represent around 80 to 85% of existing reactors. This is a thermal neutron reactor design, the modern of which is the Japanese Advanced Pressurized Water Reactor, American AP1000, Chinese Hualong and Russian VVER-1200 pressurized Reactor. Fig. **(6.10)** indicates the Diablo Canyon Power Plant, a pressurized water reactor (PWR) for electricity generation *via* a nuclear power plant near Avila Beach in San Luis Obispo County, California.

Fig. (6.10). The diablo canyon power plant. Courtesy: marya from San Luis Obispo, USA, CC BY 2.0 <https://creativecommons.org/licenses/by/2.0>, *via* Wikimedia Commons. Source: https://en.wikipedia.org/wiki/File:Diablo_canyon_nuclear_power_plant.jpg

2.8.2. Boiling Water Reactors (BWR)

A BWR is comparable with a PWR but does not use the steam generator. The lower pressure of its cooling water permits it to boil inside the pressure container, generating the steam that rounds the turbines. As in a pressurized water reactor, there is no primary and secondary twist. The thermal efficiency of these reactors can be advanced, and they can be greener, and even possibly more harmless and steady. This is a thermal neutron reactor scheme, the modern of which are the Advanced Boiling Water Reactor and the Cost-effective Streamlined Boiling Water Reactor.

Fig. (**6.11**) indicates model of the nuclear power plant from Toshiba with advanced boiling water reactor (ABWR) which is a III Generation boiling water reactor. The ABWR is currently offered by GE Hitachi Nuclear Energy (GEH) and Toshiba. The ABWR generates electrical power by using steam to power a turbine connected to a generator; the steam is boiled from water using heat generated by fission reactions within nuclear fuel.

Fig. (6.11). Model of the nuclear power plant from Toshiba with ABWR. Source: https://upload.wikimedia.org/wikipedia/commons/e/e8/ABWR_Toshiba_1.jpg. Courtesy: na0905, CC BY 2.0 <https://creativecommons.org/licenses/by/2.0>, *via* Wikimedia Commons

2.8.3. Pressurized Heavy Water Reactor (PHWR)

2.8.3.1. The CANDU Qinshan Nuclear Power Plant

A Canadian design (known as CANDU) is very similar to PWRs but uses heavy water. While heavy water is profoundly more costly than ordinary water, it creates a higher number of thermal neutrons, permitting the reactor to control without fuel supplementation. As an alternative, one can use a single large pressure vessel as in a PWR; the fuel is enclosed in hundreds of pressure tubes. These reactors are powered with regular uranium and are thermal neutron reactor designs. PHWRs can be refilled at full power and supports online refueling, making them very effective in their use. CANDU PHWRs have been built in Canada, Argentina, China, India, Pakistan, Romania, and South Korea. Fig. (**6.12**) indicates the CANDU Nuclear Power Plant in Zhejiang, China. The CANDU is a Canadian pressurized heavy water reactor design used to generate electric power. The acronym denotes its deuterium oxide moderator and its use of uranium fuel. CANDU reactors were first developed in the late 1950s and 1960s by a partnership between Atomic Energy of Canada Limited (AECL), the Hydro-Electric Power Commission of Ontario, Canadian General Electric, and other companies.

Fig. (6.12). CANDU Nuclear Power Plant at Zhejiang, China. Source: https://upload.wikimedia.org/wikipedia/commons/4/4c/CANDU_at_Qinshan.jpg. Courtesy: By Atomic Energy of Canada Limited, Attribution, https://commons.wikimedia.org/w/index.php?curid=76888406

2.8.4. Gas-cooled Reactor (GCR) and Advanced Gas-cooled Reactor (AGR)

These designs have a high thermal efficiency when matched with PWRs due to higher operative temperatures. There are a number of working reactors of this design, typically in the United Kingdom, where the concept was industrialized. Older designs (*i.e.*, Magnox stations) are either downcast or will be in the near future. Nevertheless, the AGRs have an expected life of an additional 10 to 20 years. This is a thermal neutron reactor design. Discharging costs can be high due to the large size of the reactor core. Fig. (**6.13**) indicates The Magnox Sizewell A nuclear power station. Magnox is a type of nuclear power reactor aimed to route on natural uranium with graphite as the moderator and carbon dioxide gas as the heat exchange coolant. It is appropriate to the wider class of gas-cooled reactors. The name 'Magnox' comes from the magnesium-aluminum alloy used to cover the fuel rods inside the reactor. Like most other "I Generation nuclear reactors, " the Magnox was planned with double persistence of manufacturing electrical power and plutonium-239 for Britain's nascent nuclear missiles program. The name refers precisely to the United Kingdom design but is occasionally used to refer to any similar reactor.

Fig. (6.13). The magnox sizewell: A nuclear power station. Source: https://upload.wikimedia.org/wikipedia/commons/5/51/Sizewell_A.jpg. Courtesy: CC BY-SA 3.0, https://commons.wikimedia.org/w/index.php?curid=721238

2.8.5. Liquid Metal Fast-breeder Reactor (LMFBR)

This completely unmoderated reactor design yields more fuel than it consumes. They are said to "strain" fuel because they produce fissionable fuel throughout the process because of neutron arrest. These reactors can work much like a PWR in terms of productivity and do not involve much high-pressure suppression, as the liquid metal must not be retained at high pressure, even at very high temperatures. These reactors are fast neutron, not thermal neutron projects. These reactors come in two types: (a) Lead-cooled and (b) Sodium-cooled.

2.9. Lead-cooled

liquid metal such as lead provides great radiation shielding and allows for operation at very high temperatures. Also, lead absorbs the neutrons, so fewer neutrons are lost in the coolant, due to which the coolant does not become radioactive. Unlike sodium, lead is mostly inert, so it is less prone to the risk of explosion or accident, but due to its toxicity, such large quantities of lead is difficult to dispose. Often a reactor of this type would use a lead-bismuth eutectic mixture. In this case, the bismuth would present minor radiation problems, as it is not quite as transparent to neutrons and can be transmuted to a radioactive isotope more readily than lead. The Russian Alfa class submarine uses a lead-bismuth-cooled fast reactor as its main power plant. Fig. (**6.14**) indicates Lead cooled fast reactor scheme utilized in the Alfa class submarine; the Alfa class, Soviet designation Project 705 Lira (Russian: Лира, meaning "Lyre," NATO reporting name Alfa), was a class of nuclear-powered attack submarines in service with the Soviet Navy and later with the Russian Navy. They were the fastest military submarines ever built, with only the example submarine K-222 exceeding them in submerged speed [17].

2.10. Sodium-cooled

Most LMFBRs are of this type. The TOPAZ, BN-350 and BN-600 in USSR; Superphenix in France; and Fermi-I in the United States were reactors of this type. The sodium is relatively stress-free to acquire and work with, and it also essentially stops oxidization on the numerous reactor portions submerged in it. Nevertheless, sodium bursts aggressively when open to the water, so care must be taken, but such blasts would not be more forceful than leakage of superheated fluid from a pressurized-water reactor. The Monju reactor in Japan suffered a sodium leak in 1995 and could not be resumed until May 2010. The EBR-I, the first reactor to have a core breakdown in 1955, was also a sodium-cooled reactor. Fig. (**6.15**) indicates the TOPAZ nuclear reactor. The TOPAZ nuclear reactor is a lightweight nuclear reactor developed for long-term space use by the Soviet Union.

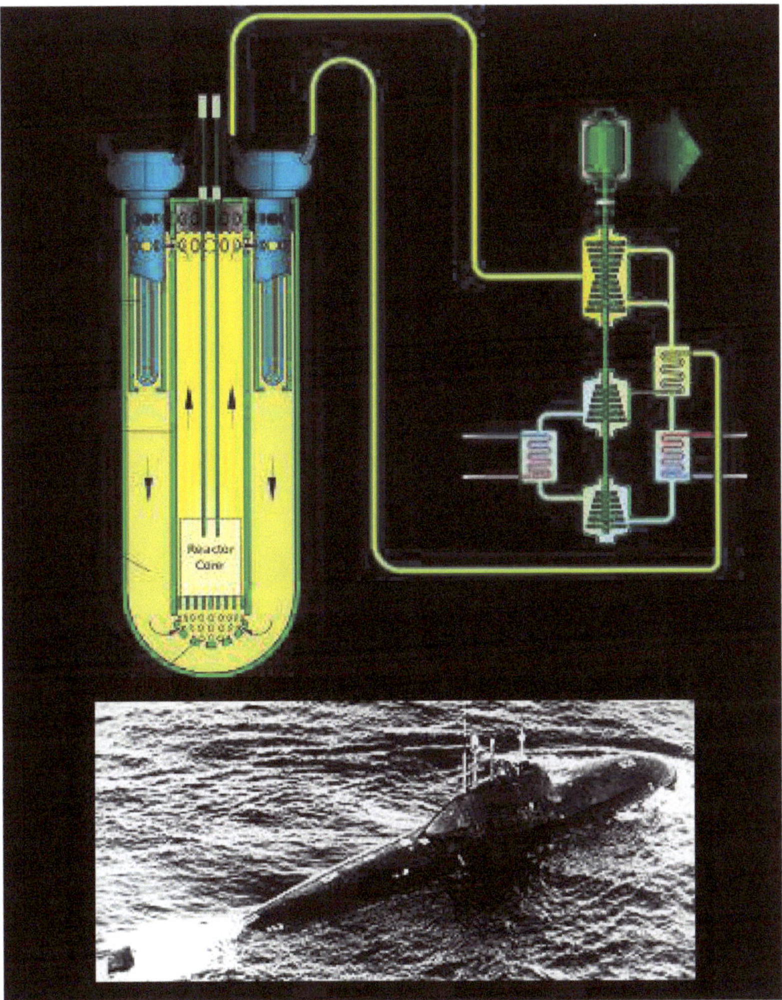

Fig. (6.14). Lead cooled fast reactor scheme utilized in Alfa class submarine. Source: https://en.wikipedia.org/wiki/File:Alfa_class_submarine_2.jpghttps://en.wikipedia.org/wiki/Lead-cooled_fast_reactor Courtesy: Soviet Military Power, 1984. Photo No. 121, page 105.

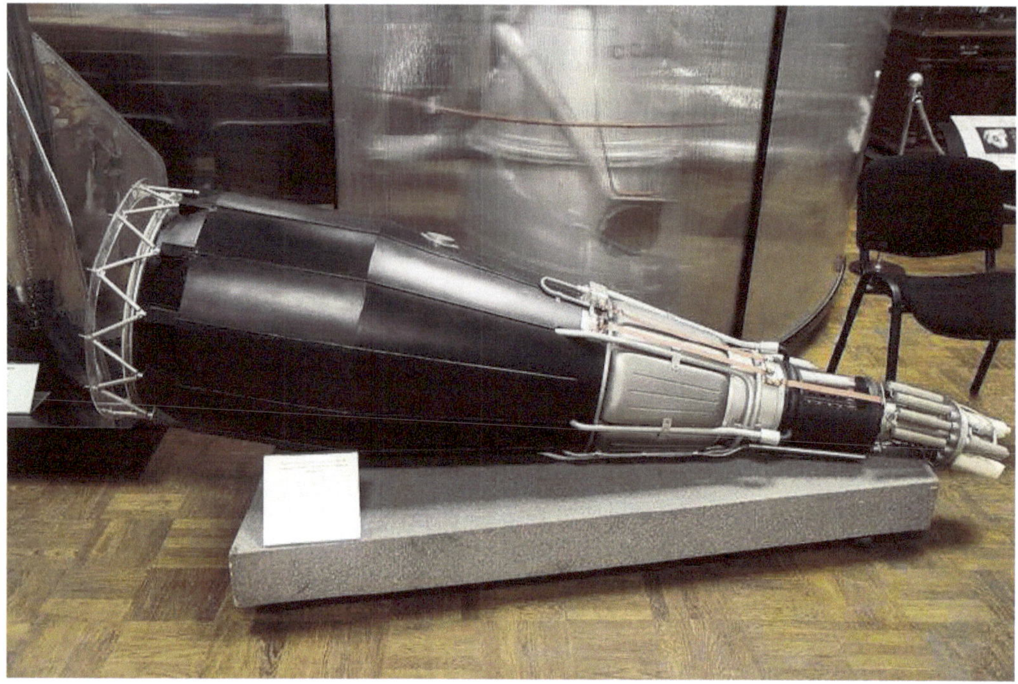

Fig. (6.15). TOPAZ nuclear reactor. Source: https://en.wikipedia.org/wiki/Nuclear_reactor#/media/ File: Topaz_nuclear_reactor.jpg. Courtesy: https://commons.wikimedia.org/w/index.php?curid=12739941

2.10.1. Pebble-bed Reactors (PBR)

These use fuel molded into ceramic balls and circulate gas through the balls. The result is a well-organized, low-maintenance, nontoxic reactor with low-cost, homogenous fuel. The model was the AVR, and the HTR-10 is working in China, where the HTR-PM is industrialized. The HTR-PM is anticipated to be the first generation IV reactor to come into operation [18].

2.10.2. Aqueous Homogeneous Reactor (AHR)

These reactors are used as fuel soluble nuclear salts (usually uranium sulfate of uranium nitrate) dissolved in water and mixed with the coolant and the moderator. As of April 2006, only five AHRs were in operation [2].

REFERENCES

[1] International Atomic Energy Agency, Nuclear Power Reactors in the World, Reference Data Series No. 2, IAEA, Vienna (2022).

[2] Ian Hore-Lacy, 3 - Nuclear Power, Nuclear Energy in the 21st Century, Academic Press, 2007, Pages 37-53, ISBN 9780123736222. https://www.sciencedirect.com/science/article/pii/B9780123736222500064 [http://dx.doi.org/10.1016/B978-012373622-2/50006-4]

[3] W. Oldekop, Electricity and Heat from Thermal Nuclear Reactors. *Primary Energy, Berlin.* Springer Berlin Heidelberg: Heidelberg, 1982, pp. 66-91. [http://dx.doi.org/10.1007/978-3-642-68444-9_5]

[4] "DOE Fundamentals Handbook: Nuclear Physics and Reactor Theory", *US Department of Energy,* 2008.

[5] D.A.R. Babu, A. Raman, P. Ashok Kumar, and D.N. Sharma. Portable Micro-R survey meters for environmental gamma radiation monitoring. Radiation Protection and Environment, 26(1-2, pt2), 400-402. 2003.

[6] M. Karimi, A.E. Rodrigues, J.A.C. Silva,10 - Biomass as a source of adsorbents for CO_2 capture, Editor(s): Mohammad Reza Rahimpour, Reza Kamali, Mohammad Amin Makarem, Mohammad Karim Dehghan Manshadi, Advances in Bioenergy and Microfluidic Applications, Elsevier, 2021, Pages 255-274, ISBN 9780128216019. [http://dx.doi.org/10.1016/B978-0-12-821601-9.00010-8]

[7] J. Bernstein, *Nuclear Weapons: What You Need to Know.* Cambridge University Press, 2008, p. 312.

[8] B.F. Towler, Chapter 7 - Nuclear Energy, Editor(s): Brian F. Towler, The Future of Energy, Academic Press, 2014, Pages 135-159, ISBN 9780128010273. [http://dx.doi.org/10.1016/B978-0-12-801027-3.00007-5]

[9] "Reactor Protection & Engineered Safety Feature Systems", *The Nuclear Tourist.*

[10] *Chernobyl: what happened and why? by CM Meyer, technical journalist,* 2013.

[11] T., Pavel; U., Shoaib (2011). Krivit, Steven (ed.). Nuclear Energy Encyclopedia: Science, Technology, and Applications. Hoboken, NJ: Wiley. pp. 48, 85. ISBN 978-0-470-89439-2.

[12] E. Nissan, "An Overview of AI Methods for in-Core Fuel Management: Tools for the Automatic Design of Nuclear Reactor Core Configurations for Fuel Reload, (Re)arranging New and Partly Spent Fuel", *Designs,* vol. 3, no. 3, p. 37, 2019. [http://dx.doi.org/10.3390/designs3030037]

[13] A. P. Meshik, "The Workings of an Ancient Nuclear Reactor", *Scientific American,* vol. 293, no. 5, pp. 82-6, 2005. [http://dx.doi.org/10.1038/scientificamerican1105-82]

[14] E. Mervine, *Nature's Nuclear Reactors: The 2-Billion-Year-Old Natural Fission Reactors in Gabon, Western Africa,* 2011.

[15] F. Gauthier-Lafaye, P. Holliger, and P.L. Blanc, "Natural fission reactors in the Franceville basin, Gabon: A review of the conditions and results of a “critical event” in a geologic system", *Geochim. Cosmochim. Acta,* vol. 60, no. 23, pp. 4831-4852, 1996. [http://dx.doi.org/10.1016/S0016-7037(96)00245-1]

[16] "World Nuclear Association Information Brief -Research Reactors". Archived from the original on 31 December 2006.

[17] "Fastest military submarine", *Guinness World Records,* 2019.

[18] "HTR-PM: Making dreams come true", *Nuclear Engineering International.*

CHAPTER 7

Radioactivity

Abstract: Radioactivity, also known as nuclear decay, radioactivity, radioactive disintegration or nuclear disintegration, is the progression by which an unstable atomic nucleus drops energy by radiation. A material comprising unbalanced nuclei is considered radioactive. Three of the most common types of decay are alpha decay (α - decay), beta decay (β-decay), and gamma decay (γ-decay). The present chapter envelops the fundamental understanding of radioactivity. It progresses through the introduction to radioactivity, alpha decay, magnetic spectrometer, determination of energy of α (Alpha) particle, Gamow's Theory of Alpha decay, beta decay, measurement of energy of beta particle and Neutrino Theory of Beta decay: (Pauli's Neutrino Hypothesis, Fermi theory of Beta decay, Gamma Decay, Measurement of Gamma γ-Ray Energies.

Keywords: Alpha Decay, Beta Decay, Gamma Decay, Radioactivity.

1. INTRODUCTION TO RADIOACTIVITY

Some nuclei are unstable and decay. These nuclei are radioactive. A nucleus can emit an alpha ray, beta ray, or a gamma-ray during its decay. Ernest Rutherford and others started studying the radiation emitted by these elements. He found three distinct forms of radiation, originally divided up based on their ability to pass through certain materials and deflection in magnetic fields. Radioactivity is a random process at the level of distinct atoms. The quantum theory proves that it is difficult to predict when a specific atom will decay, irrespective of how long the atom has been existent [1]. Nevertheless, for a majority of identical atoms, the whole decay rate can be stated as half-life or as a decay constant. The half-lives of radioactive atoms have an enormous array, from nearly instantaneous to far longer than the stage of development of the universe [2]. Radioactive decay (also known as nuclear decay, radioactivity, radioactive disintegration or nuclear disintegration) is the process by which an unbalanced atomic nucleus drops energy by particle emission. The decaying nucleus is called the parental radionuclide, and the process produces at least one daughter nuclide. Excluding gamma decay or internal conversion from a excited nuclear state, the decay is a nuclear transformation ensuing in a daughter containing an altered number of proton or

Ritesh Kohale, Sanjay J. Dhoble & Vibha Chopra

neutrons (or both). When the number of protons changes, an atom of a different chemical constituent is generated [3].

A material encompassing unbalanced nuclei is deliberated radioactive. Three of the most communal categories of decay are alpha decay (α -decay), beta decay (β-decay), and gamma decay (γ-decay), all of which include releasing one or more particles or photons. The weak force is the mechanism accountable for beta decay, while the other two are ruled by the usual electromagnetic and tough forces [4].

1.1. Alpha (α) Rays

They could barely pass through a single sheet of paper. Deflected in a magnetic field as they are positively charged particles.

1.2. Beta (β) Rays

They can pass through about 3mm of aluminium metal sheet. Deflected in the magnetic field as they are negatively charged particles.

1.3. Gamma (γ) Rays

They can pass through several centimetres of lead or concrete walls. They do not deflect in a magnetic field as they are charge-less particles.

Following Fig. (**7.1**) gives a general idea about their existence and nature.

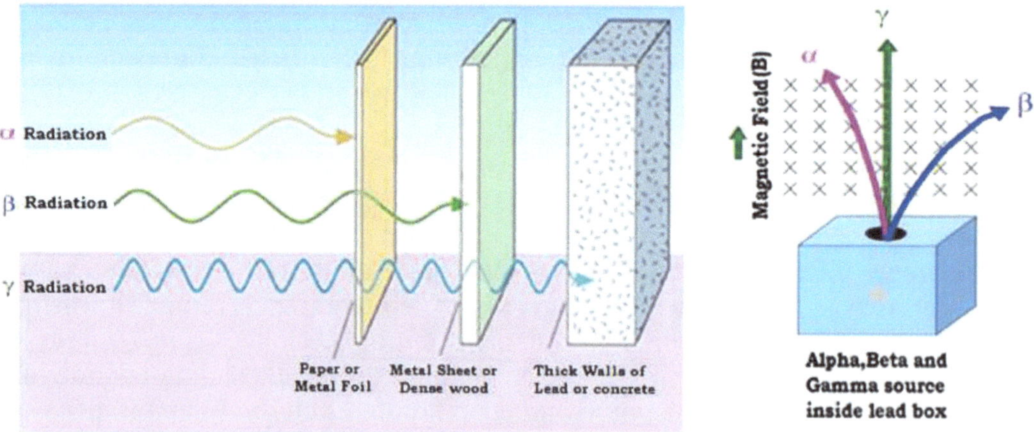

Fig. (7.1). Schematics of Alpha, Beta and Gamma rays.

2. ALPHA, BETA AND GAMMA DECAY

2.1. Origin of Alpha Decay

In alpha decay, an energetic helium ion (alpha particle) is ejected, leaving a daughter nucleus of an atomic number less than two by the parent (Z-2) and of an atomic mass number less than four by the parent (A-4). Alpha rays have been identified as helium nuclei ($^{4}_{2}He$). Fig. (**7.2**) indicates the basic mechanism of Alpha decay.

The reaction for alpha decay is,

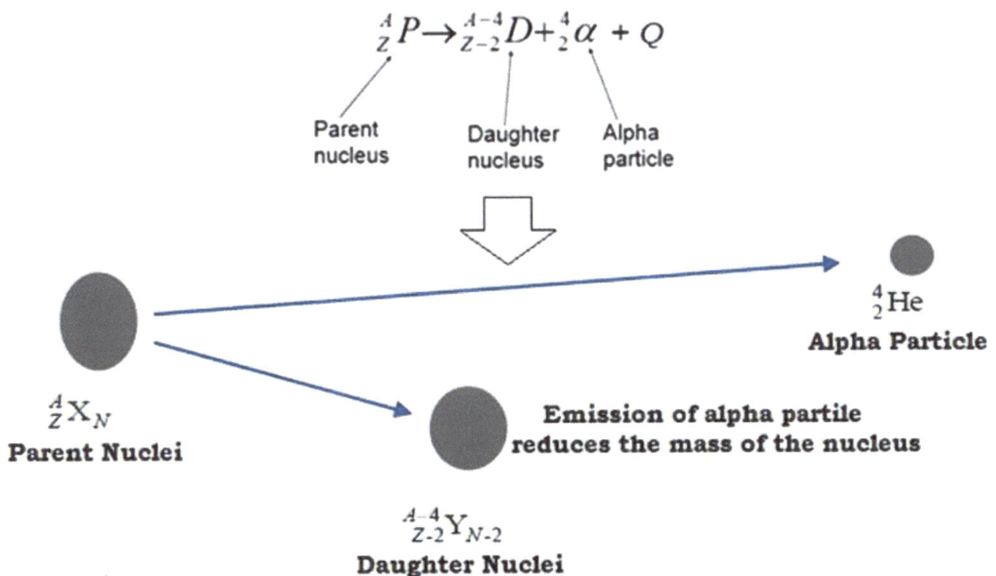

Fig. (7.2). Alpha decay.

In the process of alpha decay, the total mass of the daughter nucleus plus the alpha particle is less than the mass of the original parent nucleus.

$$\mathbf{M}_{\text{Parent}} > \mathbf{M}_{\text{Alpha}} > \mathbf{M}_{\text{Daughter}}$$

The "missing" mass is not really missing. But it has been converted into energy as per Einstein's mass-energy correlation $E = mc^2$.

The energy is found (mostly) in the kinetic energy of the alpha particle and daughter nucleus moving away from each other.

2.1.1. Properties of Alpha Rays

1. Alpha rays consist of a positively charged helium nucleus.
2. Alpha rays highly ionize the gases through which they pass.
3. Alpha rays can be deflected by electric and magnetic fields.
4. Alpha rays are electromagnetic in nature. They have a longer wavelength, so their energy is less [5].
5. Alpha rays have low penetrating power as compared to beta and gamma rays.
6. Alpha rays propagate with an average speed of $1/10^{th\ of}$ the speed of light [6].

2.1.2. Example

1. An example is the decay (symbolized by an arrow) of the abundant isotope of uranium, ^{238}U, to a thorium daughter plus an alpha particle:

$$^{238}_{92}U \rightarrow {}^{234}_{90}Th + {}^{4}_{2}He$$

Where the Q value for this α decay is Q_α=4.268 MeV

1. Alpha decay of radium-226. The daughter isotope is radon-222.

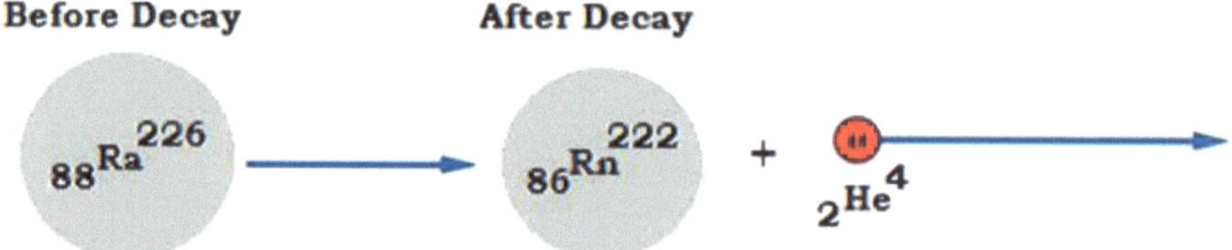

2.2. Magnetic Spectrometer

2.2.1. Determination of Energy of α (Alpha) Particle

A magnetic spectrometer is an instrument that measures the energy of an α-particle. It works on the principle of momentum selector and semicircular (180^0) magnetic focusing.

Basically, it consists of an ion source, momentum selector, transverse (perpendicular) magnetic field set up and the photographic plate. As shown in Fig. (**7.3**), the α-particles emitted from the radioactive ion source enter the chamber in the form of narrow beam through slit S. The ion source is placed at one end and photographic plate is placed at the other end. Between the source and photographic plate, the perpendicular magnetic field is applied due to which α-particles experience the force perpendicular to their path and bend their path in a circular orbit of radius 'r' [7].

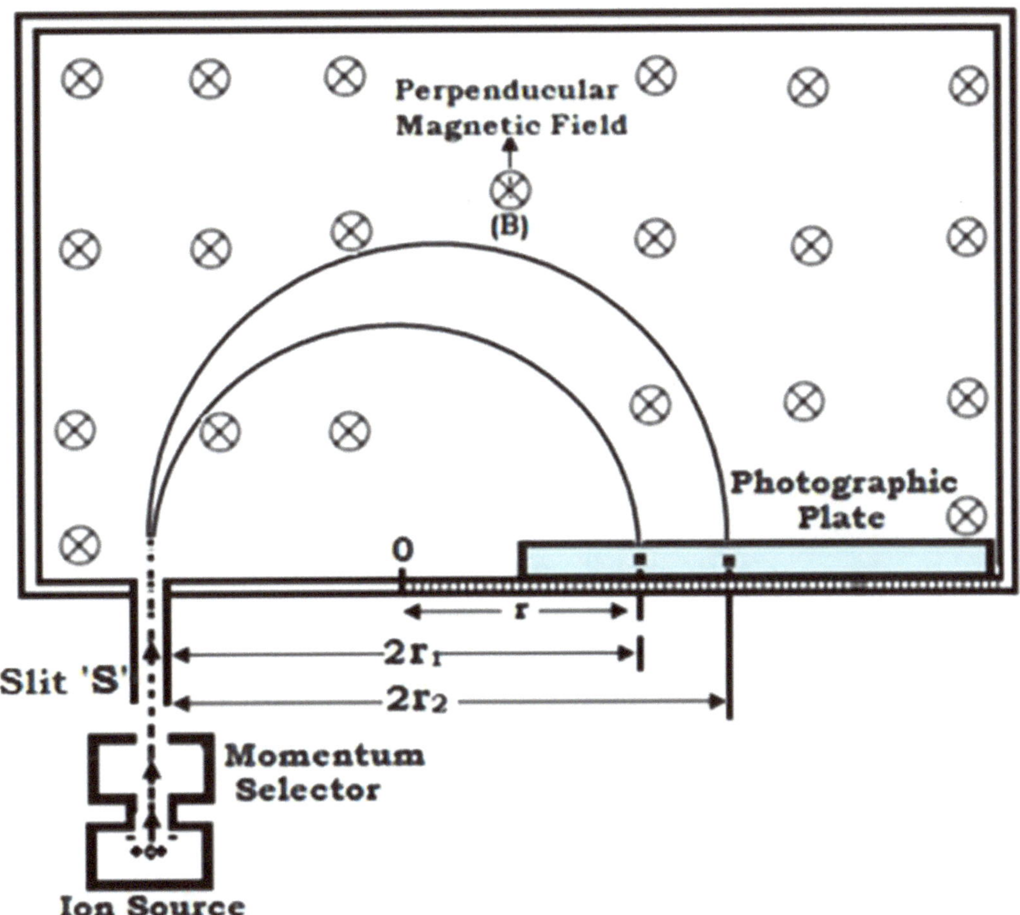

Fig. (7.3). Magnetic spectrometer for determination of the energy of α-particle.

The photographic plate is adjusted so that α-particle travelling in semi-circular path strikes the plate and produces a spot on it. Therefore the momentum and energy of the α-particle can be determined by the position of the trace on a photographic plate [8].

Under the action of perpendicular magnetic field 'B', the α-particle will describe a semi-circular path of radius 'r' given by,

$$\frac{mv^2}{r} = Bqv$$

$$\frac{mv}{r} = Bq$$

$$v = \frac{q}{m} Br \qquad (1)$$

$$r = \frac{mv}{qB}$$

Where, v, q and m are velocity, charge and mass of α-particle respectively.

From equation (1), the momentum of α-particle is,

$$P = mv$$

$$P = m \, x \, \frac{q}{m} Br$$

$$P = qBr$$

Thus, for a given magnetic field B, the α-particle of same velocity 'v' will describe a semi-circular path with radius 'r' and be focused at a photographic plate.

The kinetic energy of α-particle can be given by,

$$K.E._{alpha} = \frac{1}{2} mv^2$$

$$K.E._{alpha} = \frac{1}{2} m \left(\frac{q}{m} Br\right)^2 \qquad (2)$$

$$K.E._{alpha} = \frac{q^2m^2r^2}{2m}$$

By selecting the values B, v and r, it is observed that the value of the α-particle ranges from 4 MeV to 9 MeV.

2.2.2. Stopping Power

Nuclear physics is defined as the retarding force acting on charged particles. The energy lost per unit length of the medium remains almost constant until the α-particle loses all its energy called stopping power.

When α-particle travels through the medium, due to the interaction with matter, it loses its energy by ionization. The stopping power depends on the type of energy radiation and the material's properties through which it passes [9]. The stopping power of the material is numerically equal to the loss of energy E per unit path length x, given by:

$$S(E) = -\frac{dE}{dx}$$

The minus sign makes S positive.

2.2.3. Range of Alpha Particles

When α-particles are made to pass through the matter, it is found that they are stopped abruptly after a certain distance called as the range of α-particle (R).

The distance R in the air from the source up to which α-particle constantly travels, and after that, it comes at rest abruptly, also called range R [10].

Beyond this distance (Range 'R'), the ionizing power, response to the photographic plate and fluorescence effect disappear simultaneously.

The range of α-particle depends on:

1. Radioactive source for emission of α-particle.
2. Nature of absorbing medium or the atom density in the material.
3. Initial energy and velocity of α-particle.

2.3. Experimental Determination of Range of α-Particle

For the precise determination of the range of an α-particle, the experimental arrangement was made by Bragg called Bragg's Apparatus shown in the fig. The source 'S' which emits α-particles are placed on a movable block on one end, and an ionization chamber is placed on another end. The distance of the source from the ionization chamber can be adjusted and measured on a graduated scale. The collimating slits S_1 and S_2 are adjusted such that a narrow beam of α-particles can be produced. This narrow beam, after passing through the known thickness of air (*i.e.*, distance between source and ionization chamber) enters the ionization chamber 'C.' The ionization current produced by the α-particle at any given distance is measured by electrometer 'E' connected to the ionization chamber 'C'.

Bragg's observed that if the distance between source and ionization chamber is increased, the ionization current slowly rises to maximum up to a certain distance and then drops suddenly to zero, shown in Fig. (**7.4**). The curve relating ionization current and distance of α-particle from the source is called Bragg's Curve [11].

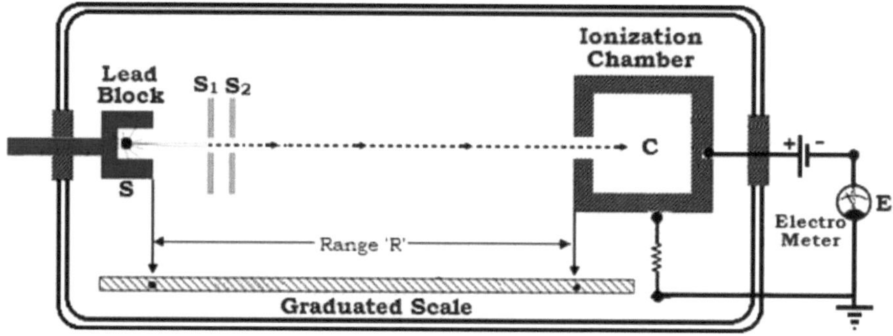

Fig. (7.4). Bragg's Apparatus for Determination of Range of α-particle.

The distance between the source and ionization chamber at which the maximum ionization current is observed is called the **range** of α-particles. Thus range is a distance at which α-particle loses all its energy in producing ionization and can be given by Geiger's Law.

Fig. (**7.5**) Indicates very rapid decay in ionization at the end of the range 'R' known as Bragg Hump, and the tail of the curve appeared due to the straggling effect.

2.3.1. Geiger's Law

Geiger Law is a correlation between the range of an α-particle ('R') and its velocity of emission ('v'), and it is observed that the range 'R' is proportional to the third power of velocity, given by:

$$R \propto v^3 \qquad \therefore R = av^3$$

Where a= constant

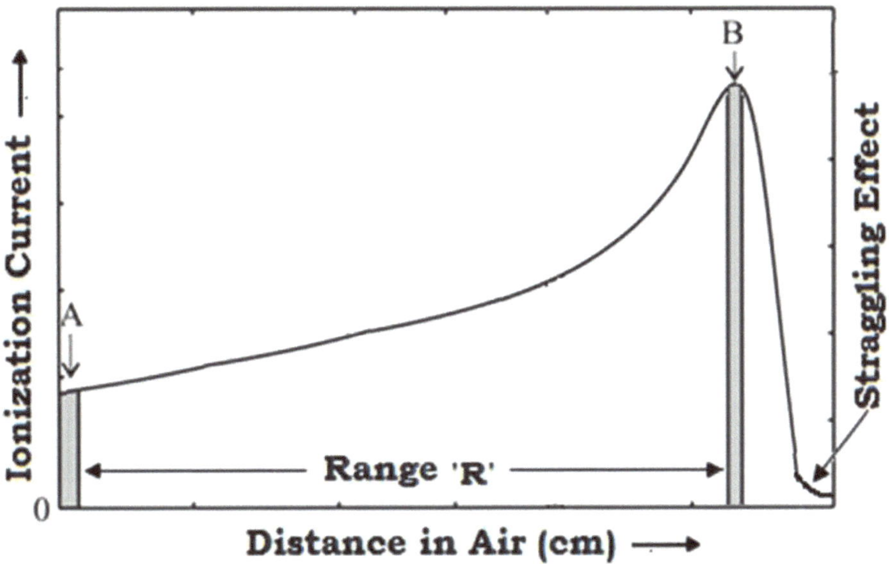

Fig. (7.5). Bragg's Curve.

2.3.2. Geiger-Nuttal Law

Geiger and Nuttal gave an empirical relation between the range 'R' of an α-particle and disintegration constant λ [12] of the radioactive element that emits an α-particle as follows:

$$\log\lambda = A + B\log R$$

Where A and B are constants

The above relation is known as Geiger-Nuttal law.

The Geiger-Nuttal law can also be given as [13]:

$$\log R = a + b\log T$$

$$\text{Where } T = \frac{0.6931}{\lambda} = \text{Half-life of an } \alpha\text{-emitter}$$

If the graph is plotted between log R and disintegration constant λ for different radioactive elements (α-emitters), then it shows a linear relationship, and a straight line is obtained for each series as shown in Fig. (**7.6**) below,

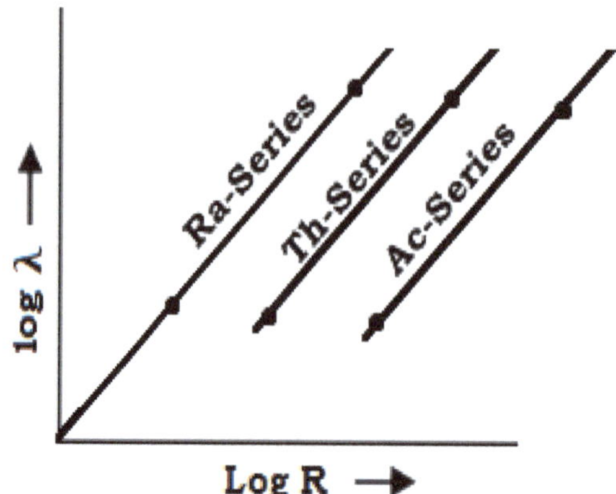

Fig. (7.6). Relation between Log R and disintegration constant λ for different radioactive elements.

2.4. Alpha Particle Tunneling

Classical mechanics expect that particles that do not have enough energy to classically overcome a barrier will not be able to reach the other side. Thus, a ball without sufficient energy to overcome the hill would roll back down. Or, lacking the energy to penetrate a wall, it would bounce back (reflection) or, in the extreme case, put in the ground itself inside the wall (absorption). In quantum mechanics, these particles can, with a very small probability, *tunnel* to the other side of the barrier, thus crossing the barrier (Fig. **7.7**). Here, the "ball" could, in a sense, *borrow* energy from its surroundings to tunnel through the wall or "roll over the hill," paying it back by making the reflected electrons more energetic [14]. This difference comes due to the matter having properties of waves and particles in quantum mechanics as well.

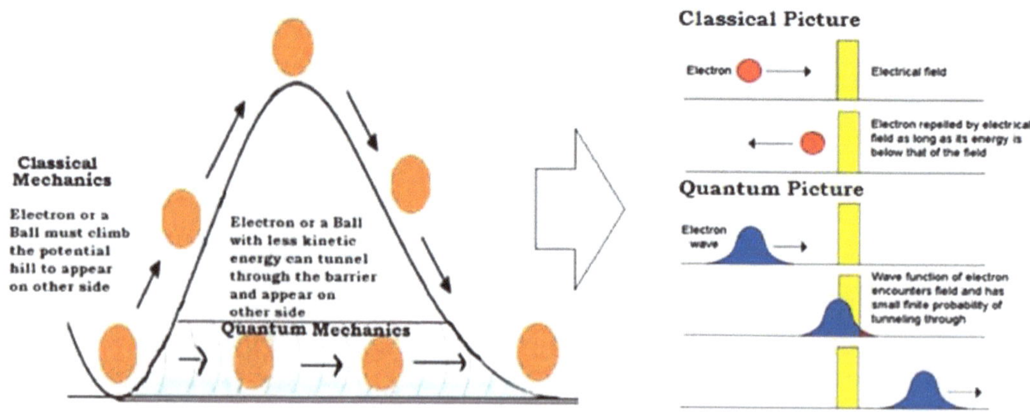

Fig. (7.7). Quantum tunneling.

Quantum Tunneling is the quantum mechanical phenomenon where a particle tunnels through a barrier that it classically cannot overcome [15].

Gamow shows a small definite probability that an alpha particle can leak through the barrier even though its kinetic energy is less than the height of the barrier. This phenomenon is known as **Alpha Particle Tunneling**.

2.5. Gamow's Theory of Alpha Decay

According to classical physics, a particle having energy E less than height V_0 of a potential barrier could not penetrate the region inside the barrier. But in quantum mechanics Gamow's Theory [16] gave a faithful explanation about the decay of α-particle that,

1. α-particle inside the heavy nucleus is considered to behave like a free particle.
2. The wave function associated with the α-particle must be continuous at the barrier will show the exponential decay inside the barrier.
3. The wave function of the α-particle must also be continuous on the other barrier, and hence there is a finite probability that the α-particle will tunnel through the potential barrier $V_{0,}$ as shown in Fig. **(7.8)**.

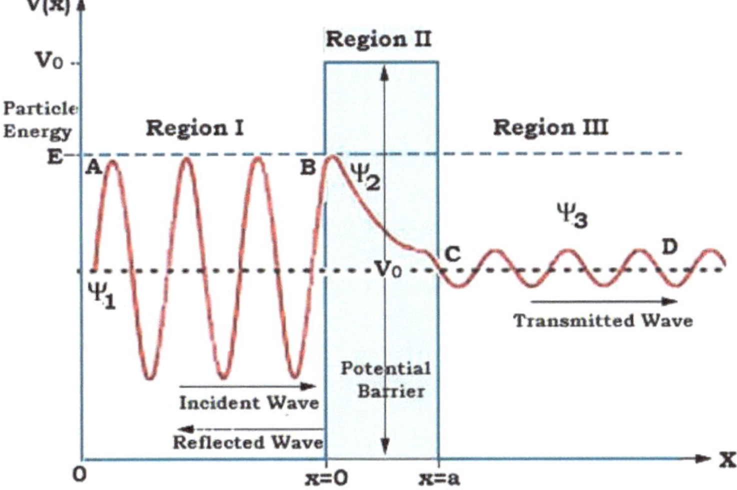

Fig. (7.8). Tunnelling of alpha particle.

Such kind of tunneling of a free particle (α-particle) is often explained in terms of Heisenberg's uncertainty principle and wave-particle duality of matter by using Schrodinger's wave equation.

During each collision of an α-particle with the walls of a nucleus, there is a certain probability P_T that the particle will penetrate through the potential barrier.

The frequency of collision of α-particle with the walls is given by,

$$f = \frac{v}{4R_n} \tag{1}$$

Where, v = velocity of an α-particle and R_n = Radius of nucleus

Thus the probability of decay of an α-particle per unit time can be given by decay constant λ as,

$$\lambda = f.\,P_T$$

$$\lambda = \frac{v}{4R_n.} P_T \tag{2}$$

Hence, if probability P_T is known, the decay constant λ can be calculated,

To calculate $P_{T,}$ consider one-dimensional coulomb's potential barrier of the rectangular shape of width 'a' and height V_0, which is greater than the kinetic energy of an α-particle such that,

$V(x) = 0$	for $0 < x$	Region I
$V(x) = V_0$	for $0 \le x \le a$	Region II
$V(x) = 0$	for $0 > x$	Region III

(Incidence, Reflection and Transmission of Alpha Particle through rectangular potential barrier)

Schrodinger's one-dimensional wave equation is given by,

$$\frac{\partial^2 \Psi}{\partial x^2} + \frac{2m}{\hbar^2}(E - V)\Psi = 0$$

$$\frac{\partial^2 \Psi}{\partial x^2} + \frac{2m}{\hbar^2}E\Psi - \frac{2m}{\hbar^2}V\Psi = 0$$

Where it is the mass of an α-particle and Ψ is the wave function

(A). For Region I

The potential V(x) = 0, and the wave function of a wave is Ψ_1,

Hence the, Schrodinger's equation can be written as,

$$\frac{\partial^2 \Psi_1}{\partial x^2} + \frac{2m}{\hbar^2} E \Psi_1 = 0$$

$$\frac{\partial^2 \Psi_1}{\partial x^2} + k_0^2 \Psi_1 = 0 \quad (3)$$

Where, $k_0^2 = \frac{2mE}{\hbar^2}$

Region I consists of both incident and reflected waves.

Hence the solution of equation (3) can be written as,

$$\Psi_1 = Ae^{ik_0x} + Be^{-ik_0x} \quad (4)$$

(B). For Region II

The potential V(x) =V_0, and the wave function of a wave is Ψ_2,

Hence, Schrodinger's equation can be written as,

$$\frac{\partial^2 \Psi_2}{\partial x^2} + \frac{2m}{\hbar^2}(E - V_0)\, \Psi_2 = 0$$

$$\frac{\partial^2 \Psi_2}{\partial x^2} - \frac{2m}{\hbar^2}(V_0 - E)\, \Psi_2 = 0 \quad (\text{Since}, V_0 > E \text{ in this region}) \quad (5)$$

$$\frac{\partial^2 \Psi_2}{\partial x^2} - k^2\, \Psi_2 = 0$$

Where, $k^2 = \frac{2m}{\hbar^2}(V_0 - E)$

In Region (II), there exists a forward-moving transmitted wave and backward moving reflected wave from the other side (wall)of the potential barrier,

Hence, the total wave function (Ψ_2) in the region (II) is given by,

$$\Psi_2 = Ce^{-kx} + De^{kx} \quad \mathbf{(6)}$$

(C). Region III

The potential V(x) = 0, and the wave function of a wave is

$$\frac{\partial^2 \Psi_3}{\partial x^2} + \frac{2m}{\hbar^2} E \Psi_3 = 0 \quad \mathbf{(7)}$$

The region (III) consists of only forward-moving transmitted waves,

Hence, the total wave function (Ψ_3) in the region (III) is given by,

$$\Psi_3 = Fe^{ik_0x} \quad \mathbf{(8)}$$

The constants A, B, C, D and F are to be determined using boundary conditions,

Boundary Conditions

Now, Ψ and $\frac{\partial \Psi}{\partial x}$ must have identical values at the boundaries.

$$\text{At } x = 0,\ \ \Psi_1 = \Psi_2 \text{ and } \frac{\partial \Psi_1}{\partial x} = \frac{\partial \Psi_2}{\partial x}$$

$$\text{At } x = a,\ \ \Psi_2 = \Psi_3 \text{ and } \frac{\partial \Psi_2}{\partial x} = \frac{\partial \Psi_3}{\partial x}$$

Now substituting the values of Ψ_1, Ψ_2 and Ψ_3 from equations (4), (6) and (8), we get

$$A + B = C + D \quad \mathbf{(9)}$$

$$ik_0A - ik_0B = -kC + kD \tag{10}$$

$$Ce^{-ka} + De^{ka} = Fe^{k_0a} \tag{11}$$

$$-kCe^{-ka} + kDe^{ka} = ik_0Fe^{k_0a} \tag{12}$$

Solving equations (9) and (12) to obtain $\frac{F}{A}$,

Assuming that barrier is thick (the width of barrier 'a' is large), we get

$$\frac{F}{A} = e^{-2ka}$$

$|\Psi_1|^2$ = The probability of particle existing in Region I and reaching B (wall of a barrier)

$|\Psi_3|^2$ = The probability of particle existing in Region III at C and beyond that.

The relative probability of an α-particle reaching from B to C in crossing the potential barrier (Transmission Probability) is given by,

$$P_T = \frac{|\Psi_3|^2}{|\Psi_1|^2}$$

To study the tunnelling effect, we are interested only in the region where there is leakage of an α-particle,

i.e., only forward-moving wave is to be considered,

So, we have:

$$\Psi_1 = Ae^{ik_0x}$$

$$\Psi_3 = Fe^{ik_0x}$$

$$\therefore\ P_T = \frac{|\Psi_3|^2}{|\Psi_1|^2} = \frac{|Fe^{ik_0x}|^2}{|Ae^{ik_0x}|^2} = \frac{|F|^2}{|A|^2} = \frac{FF^*}{AA^*}$$

Where F* and A* denote the complex conjugates of F and A

Hence, to obtain $P_{T,}$ we must get F in terms of A

Hence by assuming that the potential barrier is thick, *i.e.*, width 'a' is large and

By simplifying the above equation, we get

$$_T = e^{-2ka} = e^{-\frac{2}{\hbar}\sqrt{2m(V_0-E)}}$$

$$\text{Since, } k = \frac{\sqrt{2m(V_0-E)}}{\hbar}$$

Hence, for the actual potential barrier of width 'a', we have

$$P_T = e^{-\frac{2}{\hbar}\int_0^a \sqrt{2m(V_0-E)}\, dr} \quad \textbf{(13)}$$

Decay Constant, $\lambda = \frac{v}{4R_n} P_{T,}$ where $f = \frac{v}{4R_n}$ (from equation 1)

$$\lambda = \frac{v}{4R_n}\left(e^{-\frac{2}{\hbar}\int_0^a \sqrt{2m(V_0-E)}\, dr}\right) \quad \textbf{(14)}$$

Thus, the Alpha particle is in constant motion in the nucleus when bounded by a potential barrier. It bounces back and forth continuously from the walls of the barrier until it reaches a state of penetration or leakage.

Equation (14) gives an idea about the emission of an alpha particle and also approximates to Geiger-Nuttal Law.

3. BETA DECAY

In nuclear physics, beta decay (β-decay) is a type of radioactive decay in which a beta particle *i.e.*, fast, energetic electron or positron) is radiated from an atomic nucleus, converting the original nuclide to an isobar of that nuclide. For instance, beta decay of a neutron converts it into a proton by the emission of an electron supplemented by an antineutrino; on the other hand, a proton is transformed into a neutron by the emission of a positron with a neutrino in so-called positron emission. Neither the beta particle nor its linked (anti-) Neutrino occur within the nucleus preceding beta decay but are produced in the decay progression. With this development, unbalanced atoms acquire a steadier fraction of protons to neutrons. The possibility of a nuclide decomposing due to beta and other forms of decay is resolute by its nuclear binding energy. The binding energies of all accessible nuclides form what is called the nuclear band or ampoule of steadiness [16]. For either electron or positron emission to be dynamically conceivable, the energy discharge or Q value must be positive.

Beta rays have been identified as either electrons (β^-) or positrons (β^+).

3.1. Beta Minus Decay

In β^- decay, the weak interaction converts an atomic nucleus into a nucleus with an atomic number increased by one, while emitting an electron (e^-) and an electron antineutrino (νe). β^- decay generally occurs in neutron-rich nuclei [17].

During beta-minus decay, a neutron is converted into a proton.

$$ {}_{0}^{1}\mathrm{n} \rightarrow {}_{1}^{1}\mathrm{p} + {}_{-1}^{0}\mathrm{e} + {}_{0}^{0}\overline{\nu} $$

The reaction for beta minus decay is,

$$ {}_{Z}^{A}P \rightarrow {}_{Z+1}^{A}D + {}_{-1}^{0}\mathrm{e} + {}_{0}^{0}\overline{\nu}. $$

3.2. Beta Plus Decay

In β^+ decay or positron emission, the weak interaction transforms an atomic nucleus into a nucleus with an atomic number reduced by one while emitting a positron (e^+) and an electron neutrino (νe). β^+decay normally happens in proton-rich nuclei [18].

During beta-plus decays, a proton is converted into a neutron.

$$ {}_{1}^{1}\mathrm{p} \rightarrow {}_{0}^{1}\mathrm{n} + {}_{+1}^{0}\mathrm{e} + {}_{0}^{0}\nu $$

The reaction for beta plus decay is,

$$ {}_{Z}^{A}P \rightarrow {}_{Z-1}^{A}D + {}_{+1}^{0}\mathrm{e} + {}_{0}^{0}\nu. $$

Example: Beta decay of lead-214. The daughter isotope, bismuth-214, has a higher atomic number than lead.

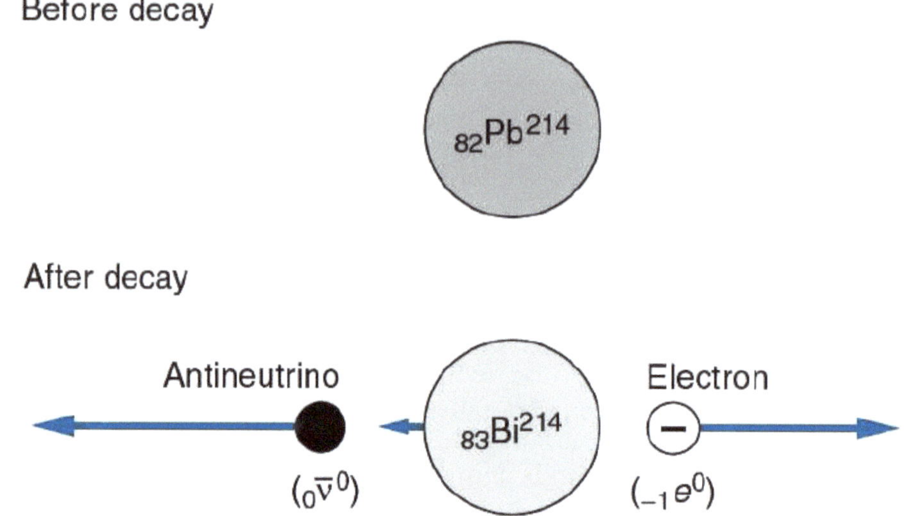

3.3. Electron or K-capture

In all cases where β^+ decay (positron emission) of a nucleus is permitted actively, so, in addition, electron capture is permissible. This is a process in the course of which a nucleus arrests one of its atomic electrons, ensuing in the emission of a neutrino [19]:

During inverse beta decay (**electron capture**), a proton in a nucleus captures an electron. The reaction is:

$$^{0}_{-1}e + ^{1}_{1}p \rightarrow ^{1}_{0}n + ^{0}_{0}\nu.$$

Energy conservation gives the Q value as,

$$Q_{EC} = (m_P - m_D)c^2$$

In this process, Q_{EC} is greater than by $2mc^2$. In electron capture, the mass of an atomic electron is converted into energy, whereas, in + decay, energy is required to create a positron. This means that EC can occur when + decay cannot. No particle is emitted in electron capture, and so, except for a very small amount of recoil energy of the daughter nucleus, the energy released escapes undetected.

3.4. Measurement of Energy of Beta Particle

The energy of the Beta (β) particle can be measured by the magnetic spectrograph. Basically, it consists of an ion source, momentum selector, transverse (perpendicular) magnetic field set up and the photographic plate. As shown in Fig. (**7.9**), the β-particles emitted from the radioactive ion source enter the chamber in the form of the narrow beam through slit S. The ion source is placed on one end, and the photographic plate is placed on the other end. Between the source and photographic plate, the perpendicular magnetic field is applied due to which β-particles experience the force perpendicular to their path and bend their path in a circular orbit of radius 'r.'

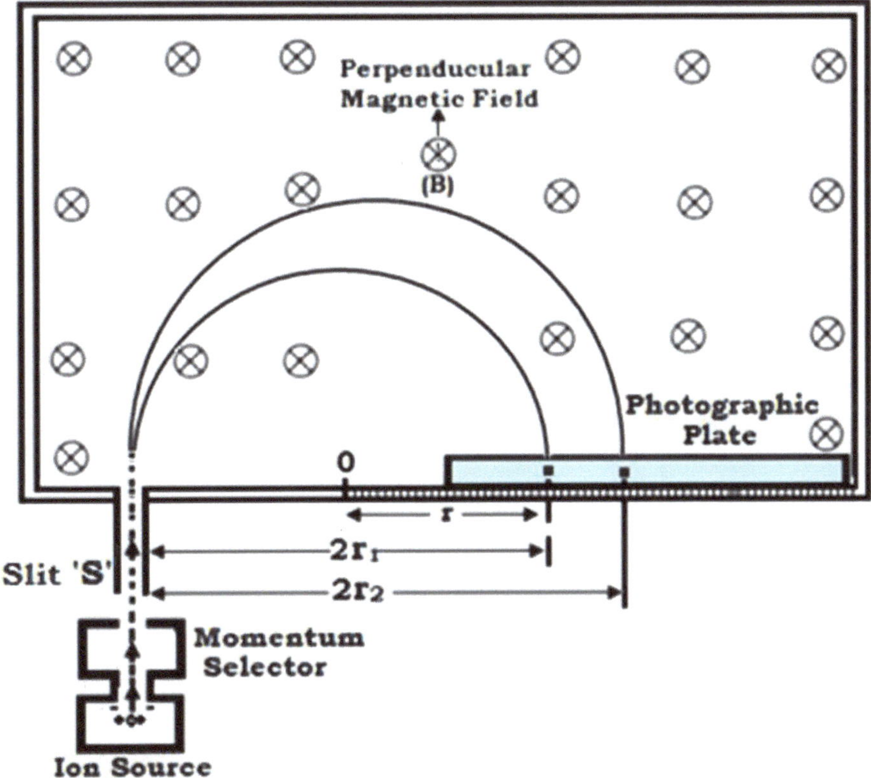

Fig. (7.9). Magnetic spectrometer for determination of the energy of β-particle.

The photographic plate is adjusted such that a β-particle traveling in a semi-circular path strikes the plate and produces a spot on it. Therefore the momentum and energy of the β-particle can be determined by the position of the trace on a photographic plate. A Geiger counter can also be used in place of a photographic plate for better results [20].

Under the action of perpendicular magnetic field 'B,' the β-particle having mass 'm' and charge 'e' will describe a semi-circular path of radius 'r' with velocity 'v' given by,

$$\frac{mv^2}{r} = Bev$$

$$\frac{mv}{r} = Be \tag{1}$$

$$mv = Ber$$

The relativistic mass of the β-particle is given by,

$$m = \frac{m_0}{\sqrt{1 - \frac{v^2}{c^2}}}$$

Where, m_0 is the rest mass of an electron

Thus putting value of m in equation (1), we get,

$$\frac{m_0 v}{\sqrt{1 - \frac{v^2}{c^2}}} = Ber$$

On solving above equation we get,

$$\frac{1}{\sqrt{1-\frac{v^2}{c^2}}} = \sqrt{1 + \left(\frac{Ber}{m_0 c}\right)^2} \tag{2}$$

Thus the kinetic energy of the β-particle can be given by,

$$K.E._{\beta} = mc^2 - m_0c^2 = \frac{m_0c^2}{\sqrt{1 - \frac{v^2}{c^2}}} - m_0c^2$$

$$= m_0c^2 \left[\frac{1}{\sqrt{1 - \frac{v^2}{c^2}}} - 1 \right]$$

$$K.E._{\beta} = m_0c^2 \sqrt{1 + \left(\frac{Ber}{m_0c}\right)^2} - 1$$

Thus, the kinetic energy of the β-particle can be determined by measuring the radius of a circular path.

3.5. Energy Spectrum of Beta (β) Particles

In the process of beta decay, either an electron or a positron is emitted. This emission is associated with the emission of **antineutrino** (β^- decay) or **Neutrino** (β+ decay), which share the energy and momentum of the decay. The beta emission has a characteristic spectrum (Fig. **7.10**). This characteristic spectrum is produced by the fact that either a neutrino or an antineutrino is emitted with the emission of a beta particle. The shape of this energy curve depends on the relative number of Beta (β) particles and kinetic energy ($K.E._{\beta}$) of Beta (β) particles. Beta particles can therefore be emitted with any kinetic energy ranging from 0 to Q value [21].

The shape of this curve first rises to a maximum and then decreases to zero value at certain kinetic energy called **end point energy.**

The end point energy is an upper limit to the energy of β-particles, which may differ for different β-emitting nuclides.

End Point Energy: The end point energy is the maximum energy with which a β-particles is emitted from a radioactive nucleus [22].

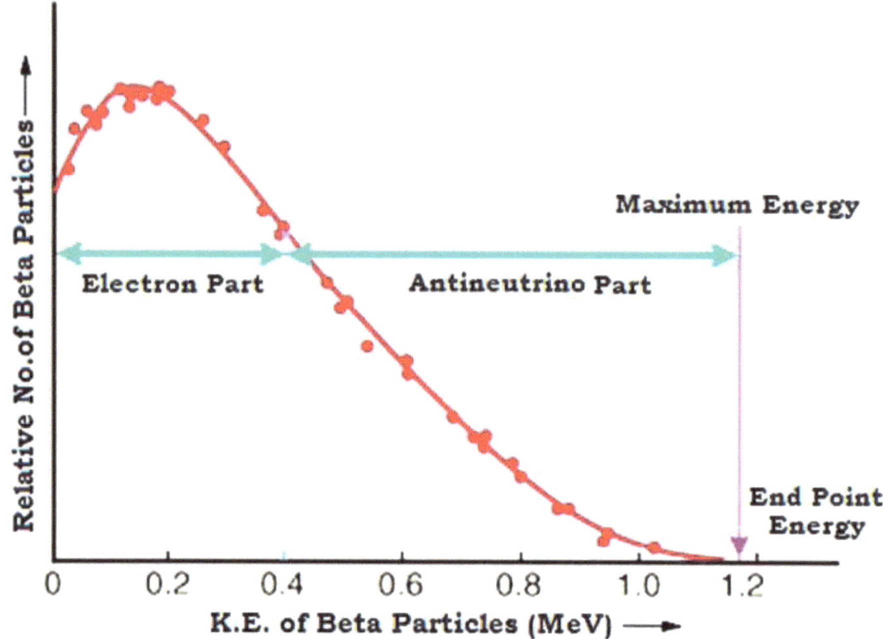

Fig. (7.10). Energy spectrum of beta particles.

3.6. Neutrino Theory of Beta Decay

3.6.1. (Pauli's Neutrino Hypothesis)

In 1930, Pauli postulated the existence of a new uncharged particle having zero rest mass, which can emit together with an electron in the process of β-decay called Neutrino.

Neutrino, explains the continuous distribution of energy of electrons in β-decay.

The Neutrino [23] was first hypothesized by Wolfgang Pauli to explain how beta decay could conserve energy, momentum, and angular momentum (spin). In contrast to Niels Bohr, who projected a statistical description of the conservation laws to clarify the experimental endless energy spectra in beta decay, Pauli postulated an unobserved particle that he called a "neutron," using the same -on ending engaged for identification of both the proton and the electron. He well-thought-out that the new particle was radiated from the nucleus together with the electron or beta particle in the progression of beta decay [24].

In accordance with the Neutrino Hypothesis, the β-decay of radioactive elements involves,

1. Conversion of proton into neutron with emission of positron and Neutrino.

$$ {}_{1}^{1}\mathrm{p} \rightarrow {}_{0}^{1}\mathrm{n} + {}_{+1}^{0}\mathrm{e} + {}_{0}^{0}\nu $$

2. Conversion of the neutron into a proton with the emission of an electron with anti-neutrino.

$$ {}_{0}^{1}\mathrm{n} \rightarrow {}_{1}^{1}\mathrm{p} + {}_{-1}^{0}\mathrm{e} + {}_{0}^{0}\overline{\nu} $$

Since the Neutrino, as well as anti-neutrino, has zero rest mass and always travels with the velocity of light and an angular spin of $\frac{1}{2}\hbar$.

3.6.2. Neutrino Properties

Neutrinos are elementary particles that have the following properties:

1. Neutrinos are electrically neutral and interact only *via* the weak interaction.
2. Neutrinos are almost massless, *i.e.*, they have very small masses compared to the other fermions.
3. Neutrinos are the most frequent particles in the Universe.
4. Neutrinos had an angular spin of $\frac{1}{2}\hbar$ same as that of an electron.
5. Neutrino has a spin vector parallel to the momentum vector, and antineutrino has a spin vector anti-parallel to its momentum vector.

3.7. Fermi Theory of Beta Decay

In particle physics, Fermi's interaction (also the Fermi theory of beta decay) clarifies the beta decay projected by Enrico Fermi in 1933 [25]. The theory postulates four fermions openly interacting with one another (at one top of the related Feynman diagram [26]). This relation clarifies the beta decay of a neutron by directly coupling a neutron with an electron, a neutrino (far ahead resolute to be an antineutrino) and a proton.

According to Pauli's Neutrino hypothesis, the existence of Neutrino can be explained based on a continuous distribution of energy of the electrons emitted in beta decay only by considering the conservation of momentum and energy of a third particle [27]. But in 1934, Enrico Fermi established a theory of beta decay that refers to the Neutrino as a massless and also chargeless particle known as Fermi's theory of Beta decay. It was the first theory for weak interactions based on quantum field theory. The interaction responsible for forming the electron and Neutrino in the beta decay is called the weak interaction. Hence, the interaction between nucleons, electrons and Neutrino can be expressed based on perturbation theory for total Hamiltonian. Fermi's Theory was the first theoretical effort to describe nuclear decay rates for β-decay.

Fig. (**7.11**) shows schematics of weak interaction, which involves the transition between nucleons (*i.e.*, protons and neutrons), electrons and Neutrino.

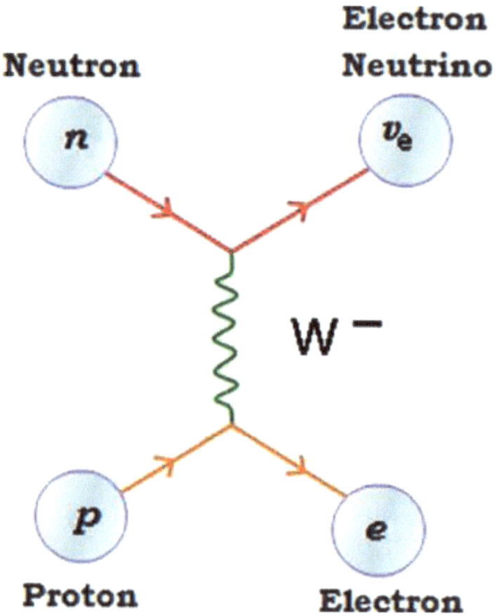

Fig. (7.11). Schematics of weak interaction, which involves the transition between protons and neutrons.

The basic assumptions of Fermi's theory were,

i. Fermi presumed that β-decay results from a particular type of interaction between the nucleons, the electron and the Neutrino.
ii. Such kinds of interactions are different from all other forces and are termed weak interactions. The strength of weak interaction can be expressed by equating it with the constant 'g.' (Where, g $=10^{-6}$ for strong interaction).
iii. The probability of decay can be expressed by the matrix element.
iv. In Beta-decay energy, E_0 is divided between electron and Neutrino.

The nature of the interaction in beta decay was given by Fermi's Golden Rule, which states that in beta decay, the transition between nucleons, electrons and Neutrino depends upon the strength of coupling between the initial and final states.'

Mathematically Fermi's Golden Rule can be expressed as follow,

$$\lambda_{if} = \frac{2\pi}{\hbar} |M_{if}|^2 \rho_f$$

Where,

λ_{if} = Transition probability,

$|M_{if}|$= Matrix element for the interaction between nucleons, electrons and Neutrino,

ρf= Density of final state.

Coupling Constant: It is a number that decides the strength of the force exerted in an interaction. The strength of the interaction in Beta decay can be determined by the coupling constant with respect to the kinetic part or between two sectors of the interaction part.

The energy released in the Beta decay process should almost all go to the kinetic energy of the β-particle.

4. GAMMA DECAY

The emission of a gamma-ray from an excited nucleus typically needs only 10^{-12} seconds. Gamma decay may also monitor nuclear reactions such as neutron capture, fission, or fusion. Gamma decay [28] is also an approach to the moderation of many excited states of atomic nuclei succeeding new categories of radioactive decay, such as beta decay, so long as these conditions retain the essential constituent of nuclear spin. When high-energy gamma rays, electrons, or protons attacks constituents, the excited atoms emit typical “secondary” gamma rays, which are yields of the formation of excited nuclear states in the attacked atoms. Such evolutions, a system of nuclear gamma fluorescence, form a matter in nuclear physics entitled gamma spectroscopy. The development of fluorescent gamma rays is a prompt subtype of radioactive gamma decay. Gamma decay is a type of radioactive decay that a nucleus can undergo. Gamma decay is different from alpha or beta decay because no particles are ejected from the nucleus when it undergoes this type of decay [29]. The electromagnetic radiation in the form of a gamma-ray photon is released. Gamma rays are photons that have very high energies and are highly ionizing. Also, gamma radiation is special in the sense that when it undergoes gamma decay, it does not change the composition or structure of the atom. During decay, it only changes the energy of the atom since the gamma-rays have no charge and are massless.

Gamma rays were determined to be high-energy photons.

A gamma-ray will be emitted when a nucleus is in an excited state when making a transition to a lower energy level.

4.1. Example

1. Gamma decay of bismuth-214. The daughter isotope is a more stable (lower-energy) version of the original bismuth-214.

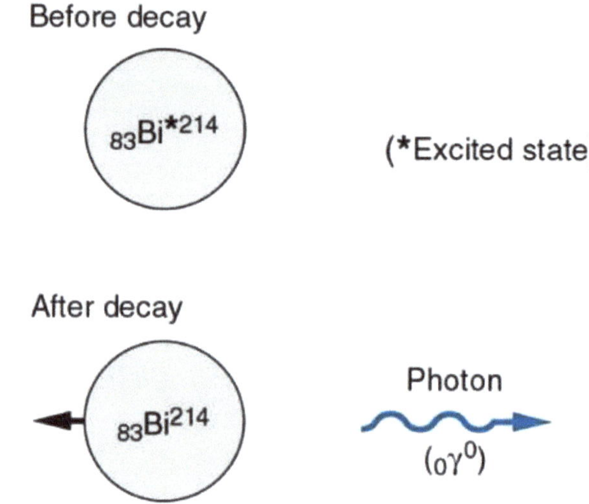

2. The γ-decay of barium-137.

$$ {}^{137}_{56}Ba \rightarrow {}^{137}_{56}Ba + {}^{0}_{0}\gamma $$

In this example, the parent atom is lowered in energy.

3. The γ-decay of plutonium-240.

$$ {}^{240}_{94}Pu \rightarrow {}^{240}_{94}B + {}^{0}_{0}\gamma $$

In this example, the parent atom is lowered in energy.

4. Decay mechanism for Cobalt-60 isotope:

Consider the decay mechanism for the Cobalt-60 isotope (Fig. **7.12**), which is most commonly used for irradiating food and as a source for radiation dosimetry.

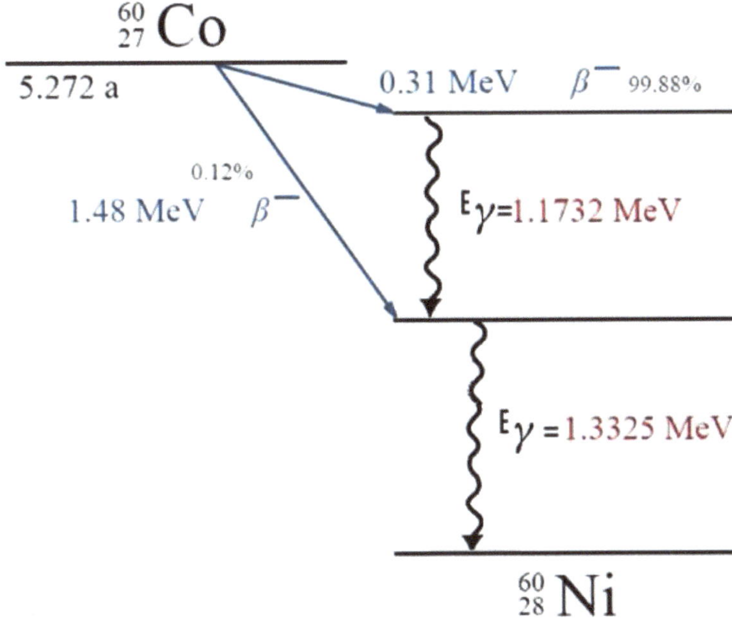

Fig. (7.12). Cobalt-60 to nickel-60 decay mechanism.

Basically, Cobalt-60 is a β-emitter source. The energy released during β-emission generates the product nucleus Nickel-60, in an excited state. When nickel-60 falls down to its ground state, it liberates γ-ray photons. Figure- shows the decay mechanism of Co-60 to Ni-60 with two strong γ-lines, 1.1732 MeV and 1.3325 MeV. This decay mechanism gives the idea that individual energy levels are present in the nucleus, and the transition between these energy levels emits the γ-ray photons [30].

4.2. Measurement of Gamma γ-Ray Energies

If a gamma spectrometer is used for recognizing samples of unidentified configuration, its energy scale must be standardized primarily. Standardization is accomplished by using the crests of a known source, such as cesium-137 or cobalt-60. Because the canal passage quantity is comparative to energy, the channel measure can then be transformed into an energy scale. If the size of the detector crystal is recognized, one can also accomplish a concentration standardization so that not only the energies but also the intensities of gamma rays or the volume of a certain isotope in the source can be determined. Because some radioactivity is present everywhere (*i.e.*, background radiation), the spectrum should be examined when not without a source. The background radiation must then be withdrawn from the real measurement. Lead absorbers can be positioned nearby the measurement device to diminish circumstantial radioactivity.

4.3. Magnetic Spectrograph for Energetic Photoelectrons (γ-ray)

The energy of γ-ray ranging from 1 MeV to 2 MeV can be determined by calculating the energy of the photo-electron emitted from the metal sample. This can be done by using a magnetic spectrograph, as shown in Fig. (**7.13**). It can also be modified with a hemispherical energy analyser and a G. M. tube for better results.

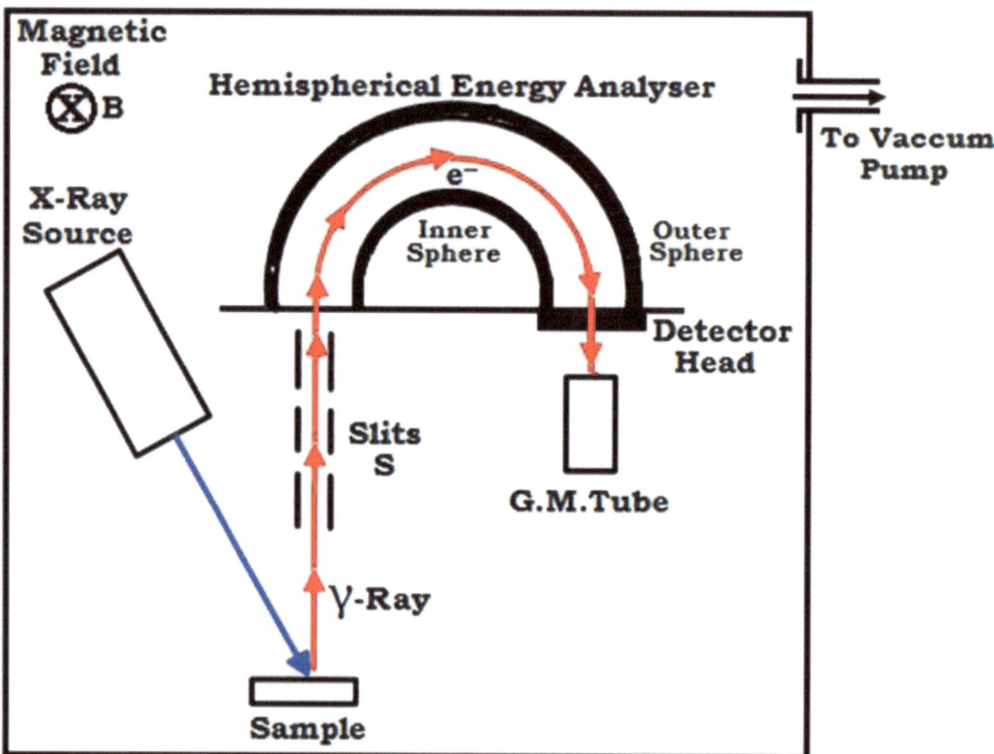

Fig. (7.13). Photo-electron energy analyser.

4.4. Detection of Energy of Photoelectrons

In this method, X-rays are made to fall on the metal sample, which emits the high-energy photo-electrons due to the photoelectric effect. These scattered photo-electrons from the sample are then made to pass through collimating slits 'S' to obtain a narrow beam. And then enters the chamber assisted with a magnetic field 'B.' The photo-electrons are then can be precisely detected by a hemispherical energy analyser and a G. M. tube [31].

Under the action of perpendicular magnetic field 'B,' these emitted photo-electrons moving with velocity 'v' will describe a semi-circular path of radius 'r,'

The centripetal force acting on an electron can be given by,

$$\frac{mv^2}{r} = Bqv$$

$$mv = Bqr$$

$$m^2v^2 = B^2q^2r^2$$

$$mv^2 = \frac{B^2q^2r^2}{m}$$

The Kinetic Energy of emitted photo-electrons is,

$$T_K = \frac{1}{2}mv^2 = \frac{B^2q^2r^2}{2m}$$

But, the Kinetic Energy of emitted photo-electron can also be given by Einstein's law of photoelectric absorption,

$$T_K = h\nu_0 - E_K$$

Where $h\nu_0$= Energy of γ-ray photon which strikes the target atom.

E_K= Binding energy of K-shell electron

Hence,

$$h\nu_0 = T_K + E_K$$

Thus by knowing the values of T_K and E_K, the energy of γ-ray can be determined.

4.5. DuMond Bent Crystal Spectrometer

It works on the principle that if a plane wave of electromagnetic radiation is scattered without loss of energy from a set of parallel planes spaced at a distance d apart, the scattered radiation is in phase if the angle of incidence equals the angle of scattering and Bragg's law, 2d sin θ = nλ is satisfied.

DuMond's experiment shows that both of these conditions can be satisfied simultaneously by a crystal that is bent in a circular arc of radius R/2. He observed that the angles of incidence and reflection would be the same at all of the reflecting planes. As the boundary surface of the crystal coincides with the focal circle, Bragg's law will be satisfied at all points on the surface when θ equals θ_B, the Bragg angle. The transmission type DuMond bent-crystal spectrometer is shown in Fig. (**7.14**).

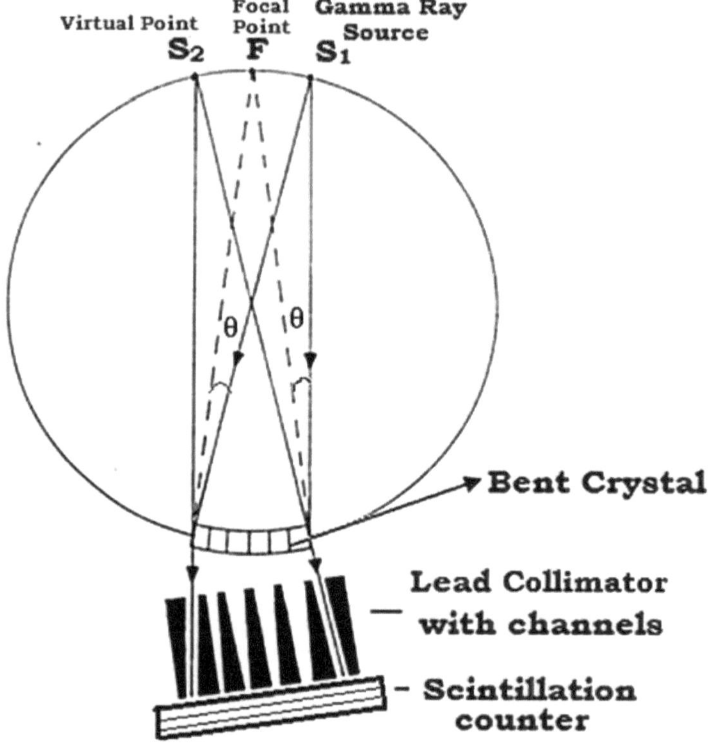

Fig. (7.14). DuMond bent crystal spectrometer.

DuMond spectrometer is concerned with the accurate measurement of gamma-ray energies and relative intensities. A bent-crystal spectrometer was used to study low-energy gamma (γ) rays from several radioactive sources having energy below 1MeV [32].

The (γ) rays from strong source S_1 are made to fall on the bent crystal, which is diffracted by the lattice planes of the bent crystal. The crystal is cut and then bent such that various diffracting planes coincide with the focal point 'F' on the diffracting circle.

Any source mounted on this diffracting circle at 'S_1' will be diffracted so that it would appear to diverge from the virtual point 'S_2'.

The lead collimator is cut and separated with channels at different angles, permitting the diffracted beam to pass into the scintillation counter for further measurements.

As we know, the scattered radiation is in phase if the angle of incidence equals the angle of scattering and Bragg's law, $2d \sin\theta = n\lambda$ is satisfied.

The counting rate can be easily determined by moving the position of the actual γ-ray source S1 on the circle. The angle at which the counting rate is maximum is the diffraction angle θ.

For first order diffraction, n=1: $2d \sin\theta=\lambda$

But,

$$\lambda = \frac{c}{\nu}$$

$$\therefore 2d \sin\theta = \frac{c}{\nu} \quad or \quad \nu = \frac{c}{2d \sin\theta}$$

Energy of Gamma Rays is,

$$E_\gamma = h\nu = \frac{hc}{2d \sin\theta}$$

Thus by knowing the values of the angle of diffraction θ, the energy of γ-rays can be determined.

Following Fig. (**7.15**) gives an idea about the actual mechanism of the DuMond bent crystal spectrometer.

4.6. Pair Spectrometer for Determination of Gamma-Ray Energy

Gamma (γ) rays having energy beyond 1.02 MeV when interacting with matter resulting in electron-positron pair formation; such phenomenons are generally used in assembling pair spectrometers. The pair spectrometer works best for measuring the energies of Gamma (γ) rays beyond 2 MeV. It is also called an electron–positron pair spectrometer because it is fabricated and planned for the immediate measurement of energy and angular correlations of e^+ and e^- pairs.

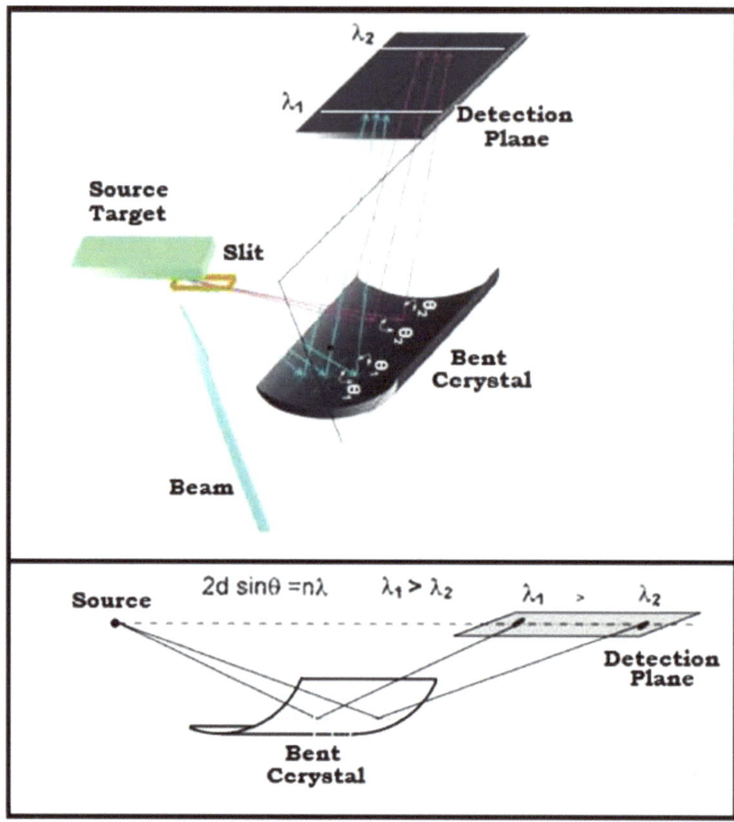

Fig. (7.15). Schematics of DuMond bent crystal spectrometer.

A typical pair spectrometer is shown in Fig. (**7.16**). Gamma rays emitted from a radioactive source pass through slit 'S' and strike the thin metallic radiator foil. Due to the interaction of gamma rays with electrons in the atoms of the target foil, electron (e^-) -positron (e^+) pairs are formed. The presence of a perpendicular magnetic field (B) bent them into a semicircular path with radius 'r' in opposite directions because of their difference in charges.

The electron and positron detectors (A and B) placed on either side of the target count the corresponding charge particles and indicate the formation of pairs [33]. The distance of detectors A and B from the target gives the diameter of the semi-circular path for electron and positron.

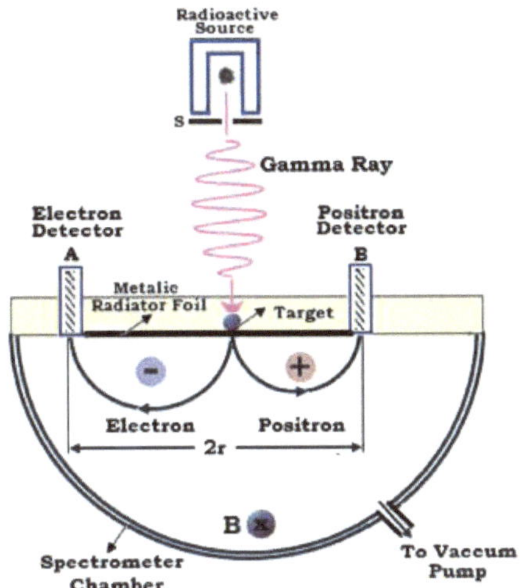

Fig. (7.16). Pair spectrometer for γ-ray energy detection.

Let E_K^+ and E_K^- be the kinetic energies of a positron (e^+) and electron (e^-) in the corresponding region, respectively,

Total kinetic energy shared by pair of two particles is given by,

$$E_K^+ + E_K^- = E_\gamma - 2m_0c^2 = E_\gamma - 1.02 \text{ MeV} \quad \textbf{(1)}$$

Where, E_γ is the energy of γ-ray,

But we know,

$$E_K^+ = P^+c \qquad \text{and} \qquad E_K^- = P^-c$$

Where, P^+ and P^- are the momentum of positron and electron, respectively?

Hence, equation (1) results as:

$$(P^+ + P^-)c = E_\gamma - 1.02 \text{ MeV}$$

Under the action of a perpendicular magnetic field, 'B' charged particle moving with velocity 'v' will describe a semi-circular path of radius 'r,'

And the centripetal force acting on a particle can be given by,

$$\frac{mv^2}{r} = Bqv$$

$$mv = Bqr$$

$$\text{or} \quad P = Bqr$$

But the momentum, $P^+ = P^- = mv$

Hence, $P^+ = Bqr^+$ and $P^- = Bqr^-$

Where, r^+ and r^- are the radii of circular paths traced by positron and electron.

$$\therefore (r^+ + r^-)Bec = E_\gamma - 1.02 \text{ MeV} \quad \textbf{(2)}$$

$$E_\gamma = (r^+ + r^-)Bec + 1.02 \text{ MeV} \quad \textbf{(3)}$$

Where,

E_γ = Energy of Gamma-ray

B = Applied magnetic field

e = Charge on electron

c = velocity of light

Using equation (3), the energy of Gamma (γ) – ray can be determined precisely.

5. APPLICATIONS OF RADIOACTIVITY

5.1. In Science

Investigation on the Earth sciences has been promoted significantly by using radiometric-dating methods, which are grounded on the standard that a specific radioisotope in environmental material degenerates at a constant identified proportion to daughter isotopes. Using such procedures, investigators have been competent to regulate the eternities of several rocks and rock formations and thereby calculate the geologic time measure. A distinct solicitation of this type of radioactivity age method involves carbon-14 dating has demonstrated especially beneficial to physical archaeologists. It has helped them to better regulate the linear categorization of past happenings by permitting them to date more precisely ancient objects and fossils from older more than 500 to 50,000 years. Radioisotopic tracers are working more precisely in ecological studies, such as, for example, those of water pollution in waterways and ponds and air pollution by smokestack emissions. They also have been used to quantify deep-water fluxes in oceans and snow-water content in watersheds. Investigators in the natural sciences, too, have made use of radioactive tracers to reveal the difficult processes. For instance, thousands of plant metabolic studies have been accompanied by compounds of sulfur, phosphorus, nitrogen and amino acids.

5.2. In Medicine

Radioisotopes have found widespread use in analysis and therapy, and this has given rise to a promptly increasing field called nuclear medicine. These radioactive isotopes have proven predominantly operative as tracers in assured diagnostic processes. As radioisotopes are indistinguishable chemically from stable isotopes of the matching element, they can take the place of the latter in biological procedures. Furthermore, because of their radioactivity, they can be readily outlined even in minute procedures with such recognition devices as relational counters and gamma-ray spectrometers. However, several radioisotopes are used as tracers; technetium-99 m, phosphorus- and 32 iodine-131 are among the most imperative. General practitioners employ iodine-131 to conclude cardiac productivity, plasma measurements, and fat metabolic rate and mainly to quantify the action of the thyroid gland where this isotope collects. Phosphorus-32 is useful in obtaining the proof of identity of malignant growths because carcinomatous cells tend to gather phosphates more than regular cells do. Technetium-99m, used with radiographic scanning devices, is appreciated for reviewing the anatomic arrangement of organs. Such radioisotopes as cesium-137 and cobalt-60 are extensively used to treat cancer. They can be controlled selectively towards malignant growths and so curtail harm to neighbouring healthy matter.

5.3. In Industry

Leading among industrial uses is power generation established on the generation of the fission energy of uranium (See Nuclear fission reactors). Additional applications comprise the use of radioisotopes to regulate and measure the width or compactness of plastic and metal sheets, to encourage polymers to cross-linking and modify plants to cultivate stronger species and preserve assured varieties of foods by killing microbes that originates decomposition. In tracer applications, radioactive isotopes are employed, for example, to identify the extent and the efficiency of motor oils to enhance the mobility of alloys for cylinder partitions and piston rings in vehicle engines [34].

REFERENCES

[1] M.G. Stabin, Radiation Protection and Dosimetry: An Introduction to Health Physics.*Radiation Protection and Dosimetry: An Introduction to Health Physics.,* M.G. Stabin, Ed., Springer, 2007. [http://dx.doi.org/10.1007/978-0-387-49983-3]

[2] L. Best, G. Rodrigues, and V. Velker, *"1.3". Radiation Oncology Primer and Review.* Demos Medical Publishing, 2013.

[3] Radioactivity: Weak Forces. *Radioactivity* EDP Sciences., 2020.

[4] W. Loveland, D. Morrissey, and G.T. Seaborg, *Modern Nuclear Chemistry.* Wiley-Interscience, 2006, p. 57.

[5] R.W. Gurney, and E.U. Condon, "Wave Mechanics and Radioactive Disintegration", *Nature,* vol. 122, no. 3073, p. 439, 1928. [http://dx.doi.org/10.1038/122439a0]

[6] a) P. Belli, R. Bernabei, and F. A. Danevich, "Experimental searches for rare alpha and beta decays", *European Physical Journal A.,* vol. 55, no. 8, pp. 1401-1407, 2019. b) "Mass spectrometer" (PDF).*IUPAC Compendium of Chemical Terminology.*, 2009. [http://dx.doi.org/10.1140/epja/i2019-12823-2]

[7] K. Siegbahn, III - Beta-ray spectrometer theory and design. Magnetic alpha-ray spectroscopy. High resolution spectroscopy, editor(s): Hai SiegbahN, Alpha-, Beta- and Gamma-Ray Spectroscopy, Elsevier, 1968, Pages 79-202, ISBN 9780720400830. [http://dx.doi.org/10.1016/B978-0-7204-0083-0.50008-0]

[8] K. Siegbahn, Alpha-, Beta- and Gamma-Ray Spectroscopy, North-Holland Publishing Co. Amsterdam 1966.

[9] K. Nakamura, "Review of Particle Physics", *J. Phys. G Nucl. Part. Phys.,* vol. 37, no. 7A, p. 075021, 2010. [http://dx.doi.org/10.1088/0954-3899/37/7A/075021]

[10] Williams, and S.C. William, *Nuclear and particle physics* Clarendon Press: Oxford, 1992.

[11] H. Geiger, and J.M. Nuttall, "XL. The ranges of the α particles from uranium", *Lond. Edinb. Dublin Philos. Mag. J. Sci.,* vol. 23, no. 135, pp. 439-445, 1912. [http://dx.doi.org/10.1080/14786440308637238]

[12] G. Gamow, "Zur Quantentheorie des Atomkernes" On the quantum theory of the atomic nucleus", *Eur. Phys. J. A,* vol. 51, no. 3-4, pp. 204-212, 1928. [http://dx.doi.org/10.1007/BF01343196]

[13] R.W. Gurney, and E.U. Condon, "Wave Mechanics and Radioactive Disintegration", *Nature,* vol. 122, no. 3073, p. 439, 1928. [http://dx.doi.org/10.1038/122439a0]

[14] M. Razavy, Quantum Theory of Tunneling. *World Scientific.* vol. 4. , 2003, p. 462. [http://dx.doi.org/10.1142/4984]

[15] "Gamow theory of alpha decay", *Archived from the original,* 2009.

[16] J. Konya, and N.M. Nagy, *Nuclear and Radio-chemistry.* Elsevier, 2012, pp. 74-75.

[17] W.D. Loveland, *Modern Nuclear Chemistry.* Wiley, 2005, p. 232. [http://dx.doi.org/10.1002/0471768626]

[18] K. Zuber, *Neutrino Physics.* 2nd ed. CRC Press, 2011, p. 466.

[http://dx.doi.org/10.1201/b11065]

[19] T. Jevremovic, *Nuclear Principles in Engineering.* Springer Science + Business Media., 2009, p. 201. [http://dx.doi.org/10.1007/978-0-387-85608-7]

[20] K. Siegbahn, Alpha-, Beta- and Gamma-Ray Spectroscopy, North-Holland Publishing Co. Amsterdam (1966).

[21] Nave, C. R. "Energy and Momentum Spectra for Beta Decay". Hyper Physics. Retrieved 2013-03-09.

[22] Balasubramanian Viswanathan, Chapter 5 - Nuclear Fission, Editor(s): Balasubramanian Viswanathan, Energy Sources, Elsevier, 2017, Pages 113-126, ISBN 9780444563538. [http://dx.doi.org/10.1016/B978-0-444-56353-8.00005-8]

[23] L.M. Brown, "The idea of the neutrino", *Phys. Today,* vol. 31, no. 9, pp. 23-28, 1978. [http://dx.doi.org/10.1063/1.2995181]

[24] F. Close, *Neutrino.* Oxford University Press, 2012. [http://dx.doi.org/10.1007/978-3-8274-2941-4]

[25] Ang, and C.N. , "Fermi's β-decay Theory", *Asia Pacific Physics Newsletter.,* vol. 1, no. 1, pp. 27-30, 2012. [http://dx.doi.org/10.1142/S2251158X12000045]

[26] D. Kaiser, "Physics and Feynman's Diagrams", *Am. Sci.,* vol. 93, no. 2, p. 156, 2005. [http://dx.doi.org/10.1511/2005.52.957]

[27] R.P. Feynman, Theory of Fundamental Processes.*Chapters 6 & 7.* W. A. Benjamin, 1962.

[28] E. Rutherford, "XV. The magnetic and electric deviation of the easily absorbed rays from radium", *Lond. Edinb. Dublin Philos. Mag. J. Sci.,* vol. 5, no. 26, pp. 177-187, 1903. [http://dx.doi.org/10.1080/14786440309462912]

[29] Ammar A. Oglat, "Gamma ray, beta and alpha particles as a sources and detection", *J. Radiat. Res. Appl. Sci,* Vol. 16, No. 1, 2023, 100503, ISSN 1687-8507. [http://dx.doi.org/10.1016/j.jrras.2022.100503]

[30] Diehl, Roland, *et al.* "SN2014J gamma rays from the 56Ni decay chain." Astronomy & Astrophysics 574 (2015): A72.

[31] G. Gilmore, and J. Hemingway, *Practical Gamma-Ray Spectrometry.* John Wiley & Sons: Chichester, 1995.

[32] G. Knoll, *Radiation Detection and Measurement.* John Wiley & Sons, Inc.: NY, 2000.

[33] Dyer, F.F., & Bate, L.C. (1972). Compilation of modern nuclear decay data for high resolution gamma spectroscopy (CONF-730302--1). United States.

[34] E.P. Steinberg, and J.O. Rasmussen, *Radioactivity.,* 2021.*Encyclopedia Britannica,* 2021.(Accessed 27 June 2021). Available from: https://www.britannica.com/science/radioactivity

CHAPTER 8

Origin and Applications of Radioactivity

Abstract: A brief introduction to radioisotopes, radiation sources, types of radiation, their applications, effects and occupational protection has been presented in this chapter. The sources of radiation (whether natural or artificial) have been discussed. However, special emphasis has been given to natural radioactive decay series and artificial radioisotopes. Applications of ionizing radiations have significantly improved the quality of human life. The contribution and application of radioisotopes in various spheres of life, *viz.* tracing, radiography, food preservation and sterilization, eradication of insects and pests, medical diagnosis and therapy and new varieties of crops in the agricultural field, have been presented briefly. In this chapter, we have first discussed the natural origin of radioactivity and the production of radioactive elements in neutron and charged-particle streams, then exchanged the data and facts to see how high-energy particles lose energy during their interaction with matter. Finally, fossil radioactivity, artificial radioactivity, applications of radioactivity in medicinal and pharmaceutical extent, and nuclear dating followed by some wide-range applications of radioactivity have been discussed.

Keywords: Artificial Radioactivity, Artificial Radioactivity, Fossil Radioactivity, Radioactivity.

1. INTRODUCTION

Radioactivity or Radioactive decay (sometimes known as radioactive disintegration, nuclear disintegration or occasionally nuclear decay) is a process by which an unstable atomic nucleus loses its energy by radiation to be more stable. A material consisting of unstable nuclei is regarded as radioactive material. The most common types of decays are alpha decay (α-decay), beta decay (β-decay), and gamma decay (γ-decay), all of which either radiate one or more particles or photons. The mechanism responsible for beta decay is the weak force, while the other two decays are attributed to the usual electromagnetic and strong forces, respectively [1]. Radioactive decay is a random process at the single atomic level. According to quantum theory, it is difficult to guess when a specific atom will decay, irrespective of the long-time existence of the atom [2]. Conversely, for a noteworthy number of indistinguishable atoms, the complete decay rate can be expressed as decay constant or half-life. The half-lives of radio-

Ritesh Kohale, Sanjay J. Dhoble & Vibha Chopra

active atoms have an enormous range that can vary from almost instantaneously to far longer than the age of the universe [3].

The decaying nucleus is termed the parent radionuclide, and this process produces at least one daughter nuclide. Excluding internal conversion or gamma decay from an excited nuclear state, the decay is a nuclear transformation ensuing in a daughter nuclide consisting of a different number of protons or neutrons (or both). An atom of a different chemical element is produced when the number of protons changes [4].

1.1. Origin of Radioactivity

Radioactive indigenous nuclides found on the Earth are deposits from the earliest supernova events that occurred before the development of the solar system. They are the elements of radionuclides that continued from that time, through the formation of the ancient solar nebula, planet mass, and up to the present. The expected short-lived radiogenic radionuclides found in today's rocks are the daughter nuclides of those primordial radioactive nuclides. Another minor source of naturally occurring radioactive nuclides is cosmogenic nuclides, formed by the attack of cosmic rays on the material in the Earth's atmosphere or shell. The decay of the radionuclides in rocks of the Earth's mantle and crust contributes to Earth's interior heat budget significantly.

Taking into account the Big Bang theory, stable isotopes of the lightest five elements (H, He, and traces of Li, Be, and B) were formed very shortly at the very beginning of the formation of the universe in a progression called Big Bang nucleosynthesis. These lightest stable nuclides such as including deuterium, still exist to date, but any radioactive isotopes of the light elements created in the Big Bang, such as tritium, have long subsequent decay. Isotopes of elements that were denser than boron were not formed at all in the Big Bang explosion, and these first five light elements do not have any long-lived radioisotopes. Therefore, all radioactive nuclei are relatively newer than the birth of the universe, having formed far along in numerous other types of nucleosynthesis in stars like supernovae, and also in the course of ongoing exchanges amongst stable isotopes and energetic particles. For example, carbon-14, a radioactive nuclide with a half-life of only 5,730 years, is continuously produced in Earth's upper atmosphere due to interactions between cosmic rays and nitrogen [5]. Nuclides formed by radioactive decay are termed radiogenic nuclides, ignoring whether they are stable or not. Stable radiogenic nuclides formed from short-lived and died-out radionuclides in the initial solar system [6]. The occurrence of other additional stable radiogenic nuclides such as xenon-129 from extinct iodine-129 in contrast to the background of stable elemental nuclides can be concluded by numerous resources.

The term radioactivity was invented by Marie Curie, who, together with her husband Pierre, began examining the occurrence in recent times revealed by Becquerel. Marie Curie and Pierre Curie removed uranium from minerals and surprisingly found that the remaining minerals exhibit more activity than the pure ones. They determined that the ore enclosed other radioactive elements. This paved the way to discover the other two elements, polonium and radium. It took four more years of treating tons of ore to segregate each constituent adequately to conclude their radioactive properties.

1.2. Fossil Radioactivity

1.2.1. Radioactive Dating of Fossils

Fossils are composed of rocks that come from identical layers. These models are widely classified and examined with a mass spectrometer. The mass spectrometer is competent in giving statistics about the nature and quantity of isotopes established in the rock. Researchers discover the proportion of parent to daughter isotope, and by associating this proportion with the parent isotope's half-life logarithmic measure, they can reveal the age of the rock or fossils.

1.2.2. Isotopes Used for Dating

A large number of joint radioactive isotopes are available, which are recycled for the dating of rocks, objects and fossils. The most common is U-235 which originated in several rocks, soil and residue. U-235 decays to Pb-207 with a half-life of 704 million years. Owing to its extended half-life, U-235 is the finest isotope for radioactive dating, predominantly of older fossils and rocks. C-14 is an alternative radioactive isotope that decays to C-12. This isotope is found in all living entities. When an animal or plant dies, the C-14 initiates decay. The half-life of C-14 is only 5,730 years. Because of its shorter half-life, the number of C-14 isotopes in a sample becomes insignificant after nearly 50,000 years, making it difficult to use for dating older samples. C-14 is used frequently in dating objects from a human being.

Maximum elements on Earth have been summarized from an astronomical mist around 4.5×10^9 years ago. This mist is typically comprised of "elemental" Hydrogen (^{1}H) and Helium (^{4}He) formed in the just first minutes after the "Big Bang." Moreover, the surrounding cloud compressed the elements produced in past generations of stars and isolated them into the intergalactic medium through numerous developments. The most radioactive nuclei were formed in supernovae, creating a combination of neutron-rich nuclei within a time-lapse of a few seconds. When, some millions of years later, the cloud compressed to form the solar system, a small number of radioactive nuclear kinds continued to exist. In the contemporary epoch, only nuclei having mean lives greater than 108 years are still present in significant abundance. The nuclei with 10^8 years < τ < 10^{12} years are listed in Table **8.1**. These long-lived nuclei consist of either highly prohibited β decays (enormous spin changes) or α-decays that happen to have Q-values that place the half-lives in this range.

Table 8.1. The chronologically arranged Nuclei with 10^8 years <$t_{1/2}$ <10^{12} indicate the "isotopic abundance" [7].

S.N.	Deacay	Half-Life (Years)	Isotopic Abundance (%)	Activity (Element) Bq kg^{-1}	Activity (Crust) Bq kg^{-1}
1	$^{40}K \rightarrow {}^{40}Ca\, e^- \bar{\nu}_e$ 89% $\rightarrow {}^{40}Ar\, \nu_e$ 11%	1.28×10^9	0.0117	3.0×10^4	6.3×10^2
2	$^{87}Rb \rightarrow {}^{87}Sr\, e^- \bar{\nu}_e$	4.75×10^{10}	27.83	8.8×10^5	8.0×10^1
3	$^{146}Sm \rightarrow {}^{142}Nd\, \alpha$	1.03×10^8	$< 10^{-7}$	< 1	$< 10^{-4}$
4	$^{147}Sm \rightarrow {}^{143}Nd\, \alpha$	1.06×10^{11}	15.1	1.3×10^5	9×10^{-1}
5	$^{176}Lu \rightarrow {}^{176}Hf\, e^-$	3.78×10^{10}	2.61	5.5×10^4	4×10^{-2}
6	$^{187}Re \rightarrow {}^{187}Os\, e^- \bar{\nu}_e$	4.15×10^{10}	62.6	1.1×10^6	8×10^{-4}
7	$^{232}Th \rightarrow {}^{228}Ra\, \alpha$	1.405×10^{10}	100	4.05×10^6	3.5×10^2
8	$^{235}U \rightarrow {}^{231}Th\, \alpha$	7.038×10^8	0.72	5.7×10^5	1.7×10^1
9	$^{238}U \rightarrow {}^{234}Th\, \alpha$	4.468×10^9	99.275	1.2×10^7	4.7×10^2

In above Table **8.1**, nuclei with 10^8 year < $t_{1/2}$ < 10^{12} years are chronologically listed. The "isotopic abundance" is for the terrestrial assortment, and the activity for the refined element resembles the terrestrial isotopic combination. For the activity in the Earth's crust, the uranium and thorium activities comprise the activities of the daughter nuclides. Reminding that three nuclides, ^{40}K, ^{147}Sm and ^{235}U, have lifetimes considerably reduced than the age of Earth (~ 4.5 $\times 10^9$ years) and therefore have precisely lesser isotopic abundances [7].

Cross-cutting interactions and distinctive superposition values permit procedures to be well-organized at a distinct site. On the other hand, they do not disclose the comparative ages of rocks conserved in two diverse extents. In this case, fossils can be convenient implements for accepting the comparative ages of rocks. Each fossil class replicates an exclusive epoch of time in Earth's olden times. The standard of faunal progression states that different fossil classes every time give the impression and vanish in the same edict, and as soon as a fossil class dies out, it vanishes and cannot rematerialize in fresher rocks as shown in Fig. (**8.1**) [8].

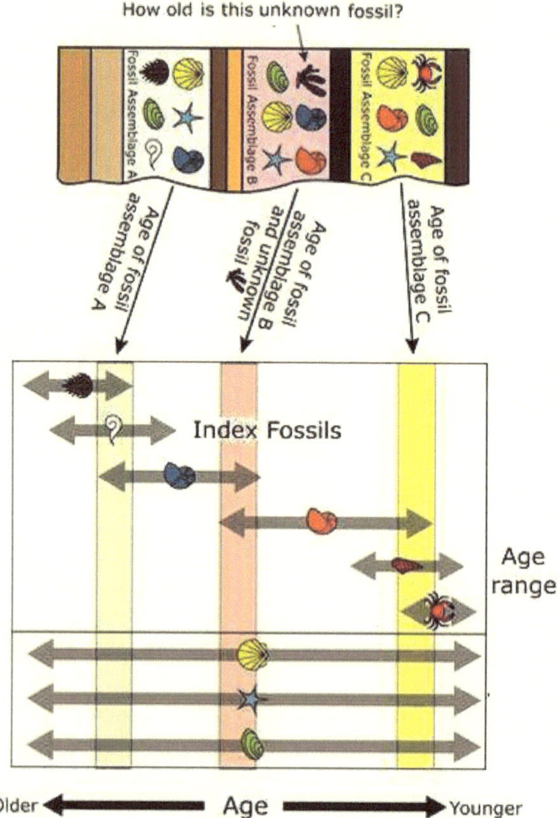

Fig. (8.1). The standard to use the fossils to recognize the relative age of fossils and rocks. (Adapted from D. J. & Deino, A. L. (2013) Dating Rocks and Fossils Using Geologic Methods. Nature Education Knowledge 4(10):1).

1.3. Half-Life

Radioisotopes decay at a continuous rate, and the time is taken to decay half the original radioisotope is acknowledged as half-life. Different radioisotopes have different half-lives and are; therefore, it is convenient to date different types of fossilized leftovers.

1.3.1. Short Range Dating

All living things enclose carbon, which occurs as a mixture of two isotopes – ^{12}C (stable) and ^{14}C (radioactive). When alive, the percentage of these two isotopes will reflect ecological stages (as carbon is continuously progressing). Once an entity expires, the proportion no longer remains permanent but modifies as ^{14}C halts downcast into ^{14}N (beta decay). Experts can measure the extent of ^{14}C residue in a model to govern how long ago it expired. ^{14}C has a half-life of only 5,730 years and so can only be applied successfully to date tasters less than ~60,000 years old [10].

1.3.2. Long Range Dating

Long-range dating can be completed by dating the rocks around the fossil to control a relative dating and age range. Dating can only be carried out on igneous rocks, not the deposited rock or fossils in which they are established [9]. Energetic volcanoes release ^{40}K from their lava, decaying into ^{40}Ar with a half-life of nearly 1.3 billion years. Using the volcanic explosion as a pointer, the age of the fossil can be estimated when a rock is deposited, or bands settled. As ^{40}Ar is inert and would have come out as a gas during the explosion, its stages in the rock designate radioactive decay. Fig. (**8.2**) reveals the radioactive isotopes and how they decay through time [11].

1.3.3. Other Dating Techniques

While radioisotope dating is the most frequently used technique for dating fossils, other techniques exist. These other practices comprise comparative dating *via* index fossils and electron spin resonance (ESR).

1.4. Index Fossils

The earth is organized into deposited layers with a grown-up section at the lowermost and new-fangled layers in the uppermost region. Different regions will not permanently have the same deposited layers owing to ecological situations such as flooding, corrosion explosions *etc.* Index fossils characterize short-lived species and thus can only be originated in a controlled gravity of rock layers. Index fossils can be used to match the ages of rock layers when other dating methods are not accessible [12].

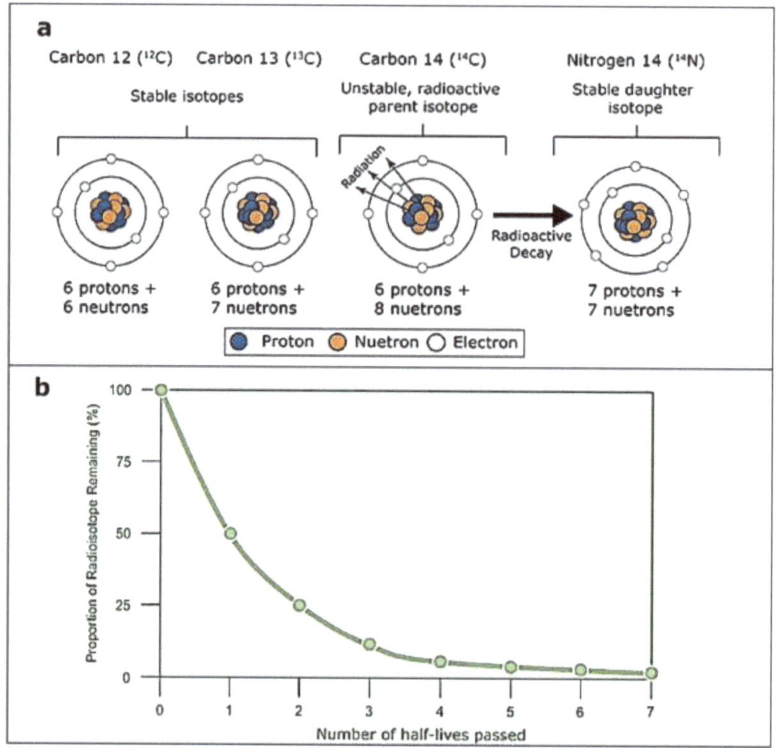

Fig. (8.2). Radioactive isotopes and how they decay through time. ***(a)*** *Carbon Dating: the atomic nucleus in C^{14} is unstable, creating the radioactive isotope. (b) The radioactive parent isotopes (atoms) into steady daughter isotopes in any mineral decay over time.*

1.5. Electron Spin Resonance

Electron spin resonance (ESR) is a convenient dating instrument to date carbon-based samples that are aged between 50,000 – 500,000 years. The information can determine ESR that when matters are buried, they are attacked by natural radiation from the soil. This originates the electrons in minerals to transfer and remain in an excited energy state. The number of high-energy electrons in an illustration can be used to conclude when the sample was buried [12].

1.6. Cosmogenic Radioactivity

Cosmogenic nuclides (or cosmogenic isotopes) are rare nuclides (isotopes) created during high-energy cosmic ray interaction with the nucleus of an *in situ* Solar System atom, causing nucleons (protons and neutrons) to be expelled from the atom, and this process is called as cosmic ray spallation.

Cosmic ray spallation, also known as the x-process, is a set of naturally arising nuclear reactions producing nucleosynthesis; it tells about the development of chemical elements after the interaction of cosmic rays on the matter. Cosmic rays are highly energetic charged particles coming from outside Earth, ranging from alpha particles, protons and nuclei of many denser elements. About 1% of cosmic rays also consist of free electrons. Cosmic rays grounds spallation when a proton-like ray particle impacts the matter, as well as other cosmic rays. As a result of the crash, large numbers of nucleons such as protons and neutrons are removed from the object smash. This development goes on not only in profound space but in Earth's higher atmosphere and upper surface (normally about ten meters) due to the continuous impact of cosmic rays. These nuclides are formed within Earth's resources such as rocks or soil, in Earth's atmosphere, and in interstellar matters such as meteorites. By calculating cosmogenic nuclides, researchers are competent to gain an understanding of various astronomical and environmental developments. There are both radioactive and stable cosmogenic nuclides. Some of these radionuclides are tritium, carbon-14 and phosphorus-32 [13]. Fig. (**8.3**) reveals an imaginative representation of the interaction and transport of cosmic rays.

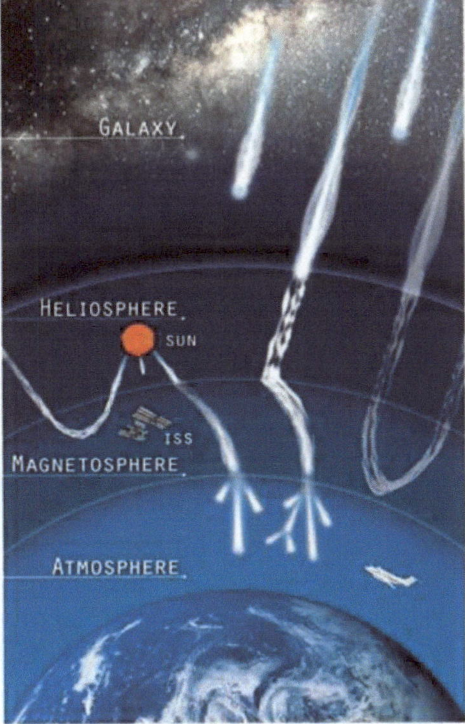

Fig. (8.3). An imaginative representation of the interaction and transport of cosmic rays [14]. (*Image Source:*https://www.researchgate.net/profile/ChristopherMertens/publication/309440486/figure/fig1/AS:86-574-0244672513@1583419967225/An-artistic-depiction-of-the-interaction-and-transport-of-cosmic-rays-through-the.ppm).

Most intergalactic material is present as a thermalized gas, typically hydrogen and helium. There occurs, nevertheless, a non-thermal constituent of cosmic rays of comparable chemical composition but with an energy spectrum (Fig. **8.4**), cresting at a kinetic energy of ~ 300MeV and then dropping like ~ E−3. Until this constituent is not completely recognized, it is assumed that it is the consequence of the motion of particles by time-varying magnetic fields created by neutron stars such as pulsars and supernova leftovers.

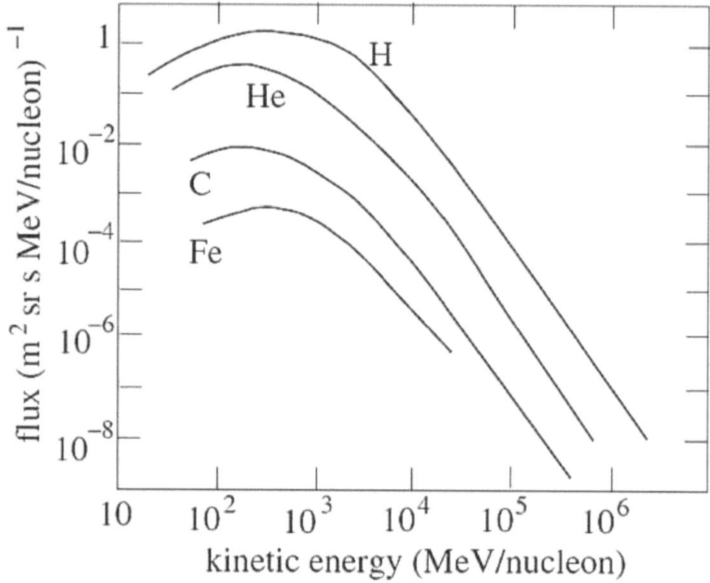

Fig. (8.4). The cosmic radiation flux outside the Earth's atmosphere [7]. Maximum particles are protons or ^{4}He nuclei with reduced numbers of heavy nuclei. Carbon and Iron are a significant case in point.

In Earth's atmosphere, cosmic rays lose their energy due to ionization and nuclear reactions since the Earth's atmosphere is adequately dense. As a result, most of the primary cosmic-radiation stopovers are in the atmosphere. Most cosmic radiations that reach the Earth's surface consist of muons and neutrinos from the decays of pions produced in these crashes. A small nuclear constituent containing mostly neutrons reaches the surface, but it gets rapidly captivated within the first few meters of the Earth's crust. Fig. (**8.5**) illustrates a "shower" prompted by a cosmic-ray proton in the upper troposphere. Two pions and a nuclear fragment are shaped when the proton strikes a nucleus. The two pions decay *via* $\pi \rightarrow \mu\nu$, and one of the muons decays *via* $\mu \rightarrow e\nu$ -ν. The undecayed muon reaches the Earth's surface, which discontinues due to its ionization energy loss. The lower pane in Fig. (**8.5**) displays the fluxes as a function of depth in the atmosphere [7].

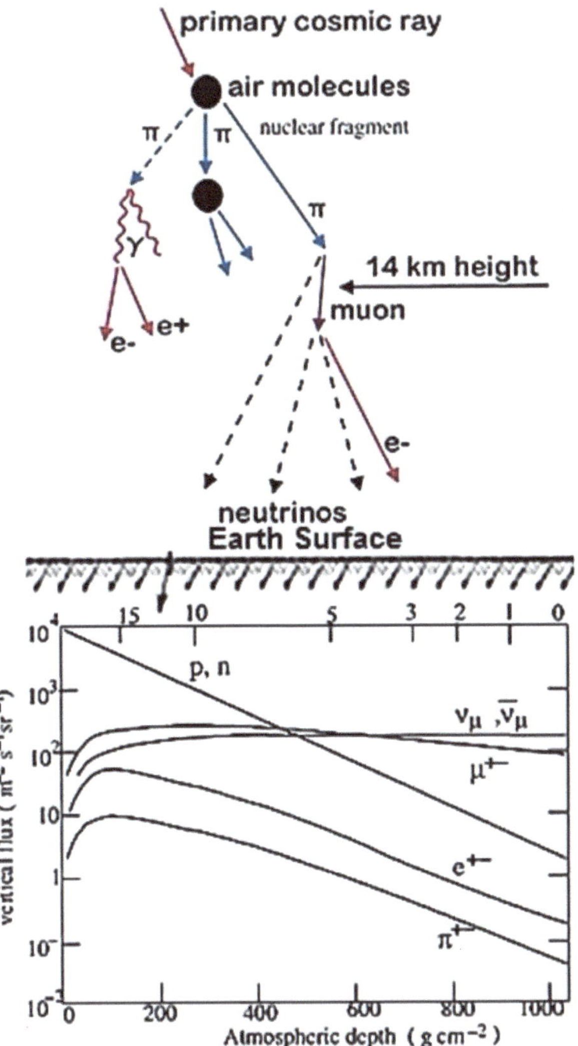

Fig. (8.5). An illustration of a "shower" prompted by a cosmic-ray proton in the upper troposphere.

In the upper atmosphere, numerous radioactive isotopes are formed when cosmic rays collide with atmospheric molecules at high speed. These isotopes are recognized as cosmogenic isotopes. The rate of production of the cosmogenic isotopes depends on the strength of the cosmic radiation, which again varies with the strength of the Earth's magnetic field and solar activity. Therefore, records of cosmogenic isotope production rates are precious to understanding the relationship between past climate change, the Earth's magnetic field, and variations in solar activity. Presently, the exact effect of past and future differences in solar activity on climate has been discussed much. The cosmogenic ice core summaries offer one of the key proceedings to resolve this argument.

1.7. Artificial Radioactivity

Artificial radioactivity, also called induced radioactivity or man-made radioactivity, is the method of altering a previously stable material into radioactive material using radiation and inducing instability [15]. The 1935 Nobel Prize in Chemistry was shared by the couple Irene Joliot-Curie and Frederic Joliot-Curie for discovering this encouraged radioactivity in 1934 [16]. Irene Curie originated her investigation with her parents, Marie Curie and Pierre Curie, after learning about the natural radioactivity in radioactive isotopes. Irene branched off from the Curies by transforming the stable isotopes into radioactive isotopes by bombarding the established material with alpha particles. The Joliot-Curies revealed that when lighter elements, such as boron and aluminum, were bombarded with α-particles, the lighter elements continued to emit radiation even after the α−source was removed or switched off. They presented that this radiation is made up of particles having one positive unit charge with a mass equal to that of an electron, now known as a positron.

Activation by a neutron is the foremost form of induced radioactivity which happens when an atomic nucleus arrests one or more free neutrons. Depending upon the chemical element involved so formed, heavier isotopes may either be stable or unstable (radioactive). Since neutrons are unstable outside the nucleus as they quickly break down within minutes outside the nucleus, only nuclear reaction, nuclear decay and high-energy interaction, such as cosmic radiation or particle accelerator emissions, can be sources of free neutrons. Neutrons that have been slowed down through a neutron moderator (thermal neutrons) are more likely to be caught by nuclei than fast neutrons.

Eliminating a neutron by photodisintegration results in a smaller amount of induced radioactivity. In its response, a gamma-ray photon hits a nucleus with an energy greater than the binding energy of the nucleus, causing the discharge of a neutron. This reaction corresponds to a minimum cutoff of 2 MeV for deuterium and approximately 10 MeV for most of the heavy nuclei [17]. Gamma rays produced by many radionuclides do not possess sufficient energy to induce this reaction. Cobalt-60 and Caesium-137 (isotopes used in food irradiation) have energy crests below this cutoff and thus terminate the stimulated radioactivity in the food [18].

Radioactivity can be induced by nuclear reactors with high neutron flux. The constituents in those reactors may be highly radioactive due to the radiation to which they are exposed. Induced radioactivity increases the amount of nuclear waste that must be disposed of ultimately, but it is not mentioned as radioactive pollution unless uncontrolled. Further exploration initially done by Irene and Frederic Joliot-Curie has led to different contemporary practices to treat various kinds of cancers [19].

Radioactive nuclei can be obtained in the laboratory by similar reactions induced by cosmic rays. However, the selection of beams and targets that takes complete benefit of production rates is in our control in the laboratory. For this, two common methods are employed, those grounded on charged particle beams and those using (thermal) neutrons produced by nuclear reactors.

The "Isotope Separator On-Line" (ISOL) and in-flight approaches are complementary in almost every respect. Using the ISOL technique, beams of high quality comparable to that of stable beams are obtained. Since we start with ions at the temperature of the target/ion source, the development is similar to the technique the beam is produced in a stable beam accelerator so one can produce beams of comparable superiority. Strong ISOL beams can be obtained, but the intensity varies noticeably according to:

a. The chemical species involved and
b. How far from stability they are

Refractory elements such as zirconium and molybdenum are enormously challenging to ionize and are thus not appropriate for this technique. This method also relies on the diffusion and effusion of the radioactive atoms present in the target, which is usually kept at high temperatures (˜2500°C) to speed up the process. Such diffusion procedures have different speeds. For short-lived nuclear species, having half-lives of the order of milliseconds or less is frequently the restraining issue in intensity since the atoms get decayed before they reach the final target. Fig. (**8.6**) schematically represents the two foremost approaches to producing a radioactive beam that has been projected. They are commonly known as the ISOL-Isotope separation online and in-flight techniques [20].

In the ISOL method, we must first create the radioactive nuclei in a target/ion source, remove them in the form of ions and accelerate them to the energy required for the experiments after selecting mass by an electromagnetic device. In contrast, the in-flight technique depends on energetic beams of heavy ions imposing on a thin target. Interfaces with the target nuclei can consequences in fission or fragmentation, with the nuclei which are shaped, leaving the target with velocities nearby to those of the projectiles.

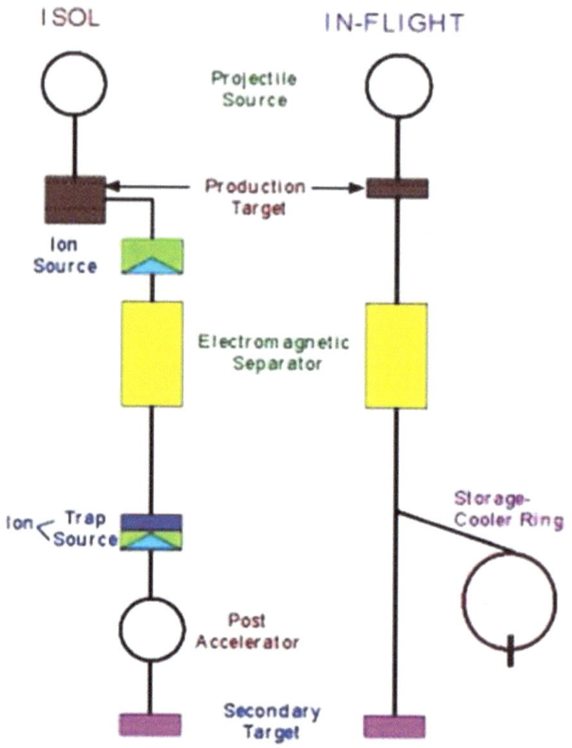

Fig. (8.6). A schematic understanding of the elementary methods of generating radioactive nuclear beams. On the left, we realize the ISOL technique with and without a post-accelerator. On the right, we realize the In-flight technique and the anticipated fusion in which fragments are trapped in a gas cell and then re-accelerated.

A combination of numerous altered types is established since the ions have high velocities and do not need additional acceleration to transport them to the secondary target. Along the target's direction, the return produced can be recognized by mass, charge and momentum in a spectrometer or fragment separator. Therefore a pure beam is not separated from the combination. As an alternative, these main features identify and acknowledge each ion, and the secondary reactions are deliberated on an event-by-event basis. A mixture of the two techniques is another option in which the in-flight reaction is transported to repose in a gas cell, drawn out and detached by mass and then re-accelerated to the necessary energy. Examining investigational and conjectural progress encompasses production, acceleration, and reactions with unstable nuclei [21].

1.8. Applications of Radioactivity

1.8.1. Medicinal and Pharmaceutical Applications

Nowadays, the radiotracer is used as an indispensable and sophisticated diagnostic tool in medicine, radiotherapy and pharmaceutical purposes.

1.8.2. Diagnostic Purpose

Technetium (99Tc) is the most communal radioactivity isotope used in the radioactive tracer. Tumours in the brain are positioned by introducing intravenously 99Tc and are then flicked through the skull with appropriate scanners. Faulty thyroid glands are studied by using ^{131}I and, most recently, ^{132}I and ^{123}I. Kidney function is also examined using a complex containing 131I.33P cast-off in DNA sequencing. Tritium (3H) is regularly used as a tracer in biological studies. The development of carbon-based fragments through metabolic pathways has been traced by using ^{14}C comprehensively.

Positron emission tomography (PET) is the most modern progress, which is the more accurate and perfect practice to find swellings in the body. A positron releasing radionuclide (*e.g.*, ^{13}N, ^{15}O, ^{18}F, *etc.*) is inserted into the patient and collected in the target muscle. A positron is released, which quickly pools with neighbouring electrons, resulting in the instantaneous release of two γ-rays in reverse directions. So produced γ-rays are noticed by a PET camera which provides a detailed symptom of their source, that is, complexity. This method is also used in imaging cardiac and brain.

The radioactive tracer yields gamma rays or single photons that a gamma camera identifies. Emissions come from altered angles, and the processer uses them to produce an image. CT scan marks the exact area of the body, like the neck or chest, or a precise organ, like the thyroid [22].

1.8.3. Therapeutic

The most common healing use of radioisotopes ^{60}Co is in the cure of cancer. During its use, leads or wrapped pointers comprising radioactive isotopes such as ^{192}Ir or ^{125}I are openly employed in the carcinogenic matter. The radiations from the radioisotopes break the tumour as long as the pointer/lead is in its place. When the conduct is complete, these are detached. This method is regularly used in treating mouth, breast, lung, and uterine cancer. ^{131}I is used for the treatment of thyroid cancers and other abnormal conditions of the thyroid. ^{32}P treats superfluous red blood cells created in the bone core.

1.8.4. Nuclear Dating

Nuclear dating, radiometric dating, radioactive dating or radioisotope dating is a procedure that is used to date constituents such as rocks or carbon, in which trace radioactive impurities are selectively combined when they are molded. The technique associates the use of certainly arising radioactive isotope within the material to the abundance of its decay products, which form at a recognized continuous rate of decay [23]. The use of radiometric dating was first put out in 1907 by Bertram Boltwood [24] and is now the major foundation of evidence about the absolute age of rocks and other environmental topographies. It can be used to date the age of fossilized life forms or of Earth itself and a wide range of natural and artificial materials.

Radiometric dating techniques composed of stratigraphic ideologies are used in geochronology to create the geologic time scale [25]. Potassium-argon dating and radiocarbon dating are among the well-known procedures. By permitting the uranium-lead dating and founding of geographical timescales, it delivers a noteworthy basis of information about the ages of fossils and the presumed proportions of evolutionary modification. Radiometric dating is also used in archaeological resources, comprising the earliest objects. Different approaches to radiometric dating are different in the timescale over which they are precise and the materials to which they can be applied. Some of the modern dating methods are listed below [26].

i. Modern dating methods
ii. Uranium-lead dating method
iii. Samarium–neodymium dating method
iv. Potassium-argon dating method
v. Rubidium–strontium dating method
vi. Uranium–thorium dating method
vii. Radiocarbon dating method
viii. Fission track dating method
ix. Chlorine-36 dating method
x. Luminescence dating methods

1.8.5. Industry and Engineering Applications

Testing of blocked water pipes and identification of leakage in oil pipes is done by radioisotopes as an industrial application. In this test, a small quantity of radioactive ^{24}Na is placed in a small bounded ball and transferred to a pipe with water. The moving ball containing the radioisotope is monitored with a detector. If the movement of the ball discontinues, it identifies the blocked pipe. Likewise, radioisotope ^{24}Na is mixed with oil flowing in a subversive pipe. With a radiation detector, the radioactivity above the pipe is examined. If there is a leakage, the radiation detector will show large movement at that specific place. Radioisotopes are also used to monitor fluid flow and detect leaks, purification and gauge locomotive wear and process erosion in apparatus.

Radioactive materials are cast-off to examine metal parts and the truthfulness of joins across various engineering and industrial applications. The titanium shell, a radioactive isotope, is positioned on one side of the object being separated, and some photographic film is placed on the other. The gamma rays pass through the object and generate an image on the film. Gamma rays identify faults in metal objects or fused junctions. This method permits critical constituents to be examined for interior faults without destruction. Radiotracer is also used to study interior imperfection without damage.

In engineering, to check the excellence of yields and to control the manufacturing procedure, production approaches must be continually monitored. The specialized maintenance is done by quality control strategies using the exclusive possessions of radiation; such devices are called nuclear gauges. They are more beneficial in dangerous chemical progression, extreme temperature, liquefied glass, and metals. The instruments are also used to measure the width of sheet materials, textiles, and paper, including metals and plastic manufacturing.

1.9. General Applications of Radioactivity

1.9.1. Agricultural Research

The thrust area of agronomic investigation includes the development of high-yielding varieties of oil seeds, plants, and other economically imperative crops and protecting plants against bugs.

1.9.2. New Varieties of Crops

Seeds of wheat, rice, maize, cotton, *etc.*, are irradiated, which undergo thoughtful genetic deviations to advance crop varieties and transformation breeding. These varieties of crops are more infection resistant and have produced high yields. Numerous republics worldwide yield a new variety of crops from radiation-induced mutants [27, 29].

1.9.3. Eradication of Insects and Pests

Sterile insect technique (SIT) is the best procedure for controlling bugs and pests. Mass-reared bugs are sterilized by irradiation so that, while they are sexually competitive, they cannot harvest progenies. As an outcome, it augments crop production and conservation of natural resources.

1.9.4. Food Preservation and Sterilization

As per WHO reports, about 25–35% of world food manufacturers are vulnerable to attack by insects, bacteria, fungi and pests, which triggers great damage to the budget of the nation. Food irradiation has more recompenses than conservative approaches. Only three types of radiation are suggested by the CODEX general standard for food irradiation which are ^{60}Co or ^{137}Cs, X-rays, or electron beams from particle accelerators. Not all types of radiation are not mentioned for food irradiation [28]. The food products are exposed to γ-radiations from the powerful well-ordered sources to kill bacteria, insects, pests, and organisms and spread shelf-life, but it also decreases the food's nutritious significance rather than terminating vitamins A, B1 (thiamin), C, and E. No radiation is leftover in the food after treatment. Radiations are categorized into three categories depending on the radiation dose and its application: they are low dose (<1 KGy), medium dose (1–10 KGy), and high dose (>10 KGy) [28].

1.9.5. Low-dose Applications (Sprout Inhibition in Bulbs and Tubers)

Irradiated potatoes can be deposited at a higher temperature of around 15°C. This not only conserves energy but also stops the sweetening of potatoes which usually arises at low temperatures. It benefits the producers of chips as low-sugar potato gives an anticipated brighter color to deep-fried potatoes and chips.

1.9.6. Delayed Ripening of Fruits

Irradiated fruits and all kinds of mangoes are irradiated with 0.25–0.75 KGy, which causes a delay in ripening practice by about 7 days, thus improving shelf-life. These doses also destroy pests. Irradiated fresh fruits can be stored for an extended period, sometimes up to 30 days at 12–14°C and in adapted atmospheres.

1.9.7. Medium-dose Applications

Seafood such as prawns and fish, fish-like mackerel, Indian salmon, Bombay duck, pomfret, and shrimp can be stored for about 7–10 days under ice. Studies have confirmed that irradiation at 1–3 KGy trailed by storing at melting ice temperatures upsurges its shelf-life nearly threefold. Meat, its products and poultry have a shelf-life of about a week at 0–3°C, which could be extended up to four weeks by applying a dose of 2–5 KGy, by deactivating spoilage bacteria.

Radiation management has improved the shelf-life of transitional moisture meat products [29].

1.9.8. High-dose Applications

While transporting the spices, they get contaminated with bug eggs and bacterial pathogens due to insufficient handling conditions. When assimilated into semi-processed or processed foods, after cooking, the microorganisms (both pathogens and spoilers) in spices can expand, which produces spoilage and thus pose a risk to customers. Several spices grow insect plague throughout storage, and unscrupulous sellers change them into spice powders. A dose of 10 KGy causes profitable infertility and recollects the natural individualities of spices. At sophisticated doses, radioactive treatment can also be applied for the overall refinement of diets for adventurous sports, low immunity patients, astronauts, military, *etc.*

REFERENCES

[1] Raymond L. Murray, Keith E. Holbert, Part II Radiation and Its Uses, Editor(s): Raymond L. Murray, Keith E. Holbert, Nuclear Energy (Eighth Edition), Butterworth-Heinemann, 2020, Page 125, ISBN 9780128128817, (https://www.sciencedirect.com/science/article/pii/B9780128128817099937). https://doi.org/10.1016/B978-0-12-812881-7.09993-7.

[2] M.G. Stabin, "3", In: *Radiation Protection and Dosimetry: An Introduction to Health Physics.*, M.G. Stabin, Ed., Springer, 2007. [http://dx.doi.org/10.1007/978-0-387-49983-3]

[3] L. Best, G. Rodrigues, and V. Velker, *"1.3". Radiation Oncology Primer and Review.* Demos Medical Publishing, 2013.

[4] W. Loveland, D. Morrissey, and G.T. Seaborg, *Modern Nuclear Chemistry.* Wiley-Interscience, 2006, p. 57.

[5] D.D. Clayton, *Principles of Stellar Evolution and Nucleosynthesis.* 2nd ed. University of Chicago Press, 1983, p. 75.

[6] B.A. Bolt, R.E. Packard, and P.B. Price, *John H. Reynolds, Physics: Berkeley.* The University of California: Berkeley, 2007.

[7] *From Nuclear Structure to Cosmology, Jean Louis Basdevant, James Rich, and Michel Spiro.* Springer: New York, 2005.

[8] Peppe, D. J. & Deino, A. L. Dating Rocks and Fossils Using Geologic Methods. Nature Education Knowledge 4(10) (2013)1.

[9] A.L. Deino, P.R. Renne, and C.C. Swisher, "40Ar/39Ar dating in paleoanthropology and archeology", *Evol. Anthropol.,* vol. 6, no. 2, pp. 63-75, 1998. [http://dx.doi.org/10.1002/(SICI)1520-6505(1998)6:2<63::AID-EVAN4>3.0.CO;2-X]

[10] G. Faure, *Mensing. T.M. Isotopes: Principles and Applications.* 3rd ed. John Wiley and Sons: New York, 2004.

[11] F.M. Gradstein, J.G. Ogg, M. Schmitz, Ed., *The Geologic Time Scale 2012, 2-volume set.* Elsevier: Waltham, MA, 2012.

[12] M. Walker, *Quaternary Dating Methods.* John Wiley and Sons: New York, 2005.

[13] A. Coc, P. Descouvemont, K.A. Olive, J-P. Uzan, and E Vangioni, "Variation of fundamental constants and the role of A = 5 and A = 8 nuclei on primordial nucleosynthesis", *Phys. Rev. D Part. Fields Gravit. Cosmol.,* vol. 86, no. 4, p. 043529, 2012. [http://dx.doi.org/10.1103/PhysRevD.86.043529]

[14] C.J. Mertens, "Overview of the Radiation Dosimetry Experiment (RaD-X) flight mission", *Space Weather,* vol. 14, no. 11, pp. 921-934, 2016. [http://dx.doi.org/10.1002/2016SW001399] [PMID: 33442336]

[15] A. Fassò, M. Silari, and L. Ulrici, "October 1999). Predicting Induced Radioactivity at High Energy Accelerators (PDF)", *Ninth International Conference on Radiation Shielding* Tsukuba, Japan.

[16] "Irène Joliot-Curie: Biographical". The Nobel Prize. n.d. Retrieved December 10, 2018.

[17] B. Thomadsen, R. Nath, F.B. Bateman, J. Farr, C. Glisson, M.K. Islam, T. LaFrance, M.E. Moore, and G. Xu, "Potential Hazard Due to Induced Radioactivity Secondary to Radiotherapy", *Health Physics,* vol. 107, no. 5, pp. 442-460, 2014. [http://dx.doi.org/10.1097/HP.0000000000000139]

[18] R.B. James, T.E. Schlesinger, Jim Lund, Michael Schieber, CHAPTER 9 - Cd1−xZnxTe Spectrometers for Gamma and X-Ray Applications, Editor(s): T.E. Schlesinger, Ralph B. James, Semiconductors and Semimetals, Elsevier, Volume 43, 1995, Pages 335-381, ISSN 0080-8784, ISBN 9780127521435, https://doi.org/10.1016/S0080-8784(08)62748-9. (https://www.sciencedirect.com/science/article/pii/S0080878408627489)

[19] Leone and Robotti. "Frédéric Joliot, Irène Curie and the early history of the positron (1932–33)", *Eur. J. Phys.* 31 975, 2010.

[20] A Carlos, Bertulani, Physics of Radioactive Beams, Chapter 1,Production of secondary beams of rare isotopes, Texas A&M University-Commerce, TX 75429, USA 2010.

[21] "Ionizing & Non-Ionizing Radiation", *Radiation Protection,* 2014.

[22] AK Sahrama, *Indian Nuclear Society Publications.,* 2010.

[23] IUPAC, Compendium of Chemical Terminology, 2nd ed. (the "Gold Book") (1997). Online corrected version: (2006–) "radioactive dating". doi:10.1351/goldbook.R05082 https://en.wikipedia.org/wiki/Chronological_dating

[24] Bertram Boltwood, "The Ultimate Disintegration Products of the Radio-active Elements. Part II. The disintegration products of uranium", *Am. J. Sci,* vol. 23, no. 134, pp. 77-88, 1907. [http://dx.doi.org/10.2475/ajs.s4-23.134.78]

[25] A. McRae, *Radiometric Dating and the Geological Time Scale: Circular Reasoning or Reliable Tools?* Radiometric Dating and the Geological Time Scale TalkOrigins Archive, 1998.

[26] J. Bernard-Griffiths, and G. Groan, "The samarium–neodymium method", In: *Nuclear Methods of Dating.,* E. Roth, B. Poty, Eds., Springer Netherlands, 1989, pp. 53-72.

[27] Bjorn Wahlstrom, "Radiation, Health and Society", *International Atomic Energy Agency,* 1997.

[28] T.D. Luckey, and K.S. Lawrence, "Radiation hormesis: the good, the bad, and the ugly", *Dose Response,* vol. 4, no. 3, p. dose-response.0, 2006. [http://dx.doi.org/10.2203/dose-response.06-102.Luckey] [PMID: 18648595]

[29] Dhir, B. Effective Removal of Radioactive Waste from Environment Using Plants. In Phytoremediation for Environmental Sustainability (pp. 71–82). 2021. Springer Nature Singapore. https://doi.org/10.1007/978-981-16-5621-7_4

CHAPTER 9

Nuclear Cosmology and Elementary Particles

Abstract: Physical cosmology is a branch of cosmology concerned with studying cosmological models. An astrophysical model, or merely cosmology, explains the largest-scale assemblies and dynamics of the universe and allows the study of necessary inquiries about its beginning, configuration, progress and conclusive fortune. In the present chapter, we tried to envelop the fundamental conceptions of Nuclear Cosmology and Elementary Particles encompassing primary and secondary Cosmic Rays (CR), the composition of CRs in the solar system and the Galaxy, elementary particles and their entire characteristics, matter and antimatter, generations of matter and fundamental forces with necessary schematics and illustrations.

Keywords: Elementary Particles, Fundamental Forces, Nuclear Cosmology, Solar System.

1. INTRODUCTION TO NUCLEAR COSMOLOGY AND ELEMENTARY PARTICLES

Cosmology is a science initiated with the Copernican principle, which suggests that heavenly bodies follow indistinguishable physical laws from those on Earth. Newtonian mechanics first permitted those physical laws to be implicit. Physical cosmology, as it is now assumed, initiated with the progress in 1915 of Albert Einstein's general theory of relativity, trailed by chief observational findings in the 1920s: first, Edwin Hubble exposed that the universe comprises an enormous number of exterior galaxies away from the Milky Way; then, work by Vesto Slipher and others presented that the universe is increasing [1]. These improvements made it conceivable to take a chance at the beginning of the universe and permitted the creation of the Big Bang theory by Georges Lemaître as the principal astrophysical model. Rare investigators still support a few unconventional cosmologies [2]; however, most cosmologists understand that the Big Bang theory gives the best details of these observations.

In nuclear particle physics and cosmology, it has been revealed that we are investigating the facts from the tiniest particles and nuclei up to the infinite universe. Moreover, fascinating enough, the study of the microscopic world, such

Ritesh Kohale, Sanjay J. Dhoble & Vibha Chopra

as particles and nuclei, is associated with the study of the macroscopic world, such as the universe. Both studies harmonize each other. It is whispered that the universe just after the big bang has incredible energy concentration and marinas novel particles unidentified to manhood. Considering the preliminary void, the existing universe has many components. If we revise the nuclear properties, we can recognize how and where these components were made in the universe. To appreciate passionate occurrences such as supernova outbursts, we need those nuclear belongings. The development and growth of galaxies can also be assumed with numerical simulation. In short, cosmology and nuclear particle physics attempt to make a widespread understanding of the history and contemporary singularities of the universe through the learning of the microscopic realm of particles and nuclei [3].

Cosmology draws profoundly on the work of numerous distinct areas of investigation in theoretical and practical physics. Areas applicable to cosmology involve particle physics research, theory, hypothetical, observational astrophysics, general relativity, quantum mechanics, and plasma physics [4].

Particle physics (similarly identified as high energy physics) is a branch of physics that reviews the natural surroundings and the particles that are set up with matter and radiation. Even though the word particle can talk about countless categories of very small entities (*e.g.*, protons, gas particles, or even domestic dust), particle physics typically examines the irreducibly tiniest noticeable particles and the essential interactions essential to designate their activities.

In particle physics, an elementary particle is a subatomic particle with no substructure, *i.e.*, it is not composed of other particles [5]. Particles at present supposed to be elementary encompass the basic fermions (quarks, leptons, antiquarks, and antileptons), which commonly are "matter particles" and "antimatter particles," as well as the gauge bosons and the Higgs boson, which normally are "force particles" that intermediate interfaces among fermions [5]. A particle encompassing two or more elementary particles is called a composite particle. In present consideration, the elementary particles are excitations of the quantum fields that similarly govern their impacts. At the moment, elementary theory clarifying these dynamic fields and particles, along with their dynamic forces, is called the Standard Model [6]. Consequently, existing particle physics examines the Standard Model and its various conceivable postponements, *e.g.*, to the latest "known" particle, the Higgs boson, or even from the past known force field, gravity [7].

1.1. Cosmic Rays (CR)

They are the artifact of cosmological reactions and breakdowns: the foremost foundations of CRs up to the knees are enormous solar radiations. In 1900: electroscopes discharged even far from normal radioactivity sources stimulated using a pole, and the ionization was dignified from the degree to which the abundances got collected due to outflow currents related to ionization. Hess 1912 (1936 Nobel prize) and Kolhörster 1914 functioned hot-air balloon which climbs up to 5-9 km and perceived that the average ionization intensifies with elevation 'A radiation of precisely extraordinary penetrating power comes into our troposphere from above.'

Thereafter Cosmic Rays were entitled by Millikan (1925) that restrained subaquatic ionization domains (he witnessed the more probing 'muon' element, not the electromagnetic as in the sky). After 1929 it was understood they are made of charged particles getting at the Earth in clusters, *i.e.*, atmospheric cascades [8].

Remarkable balloons above the height of sea level, rockets, and satellites are cast-off to scrutinize the CRs in the solar system. In forming these 'primary' CRs at the borderline of the Earth's atmosphere (>40 km), there is a constituent of the cosmological source, but then again, the main constituent overhead 1 GeV directly reaches us from the stellar space. There are decent explanations to accept as true that CRs are molded in the Galaxy apart from those with E > 1017 eV (but their impact on the energy concentration and change is insignificant).

The direct extent of cosmic rays, particularly at lower energies, has been probable since the introduction of the first satellites in the late 1950s. Particle detectors comparable to those recycled in nuclear and high-energy physics are used on satellites and interstellar investigations for research into cosmic rays. Data from the Fermi Space Telescope (2013) [9] have been inferred as confirmation that a noteworthy portion of primary cosmic rays originates from the supernova outbursts of stars [10]. Grounded on clarifications of gamma rays and neutrinos from blazar TXS 0506+056 in 2018, energetic and enormous nuclei also give the idea to produce cosmic rays [11].

Showers of high-energy particles arise when energetic cosmic rays attack the top of the Earth's troposphere. Utmost cosmic rays are atomic nuclei: most are hydrogen nuclei, some are helium nuclei, and the rest heavier elements. Even though many of the low-energy cosmic rays originate from the Sun, the backgrounds of the highest energy cosmic rays leftovers are unidentified and a matter of ample exploration. Fig. (**9.1**) illustrates air showers from very high-energy cosmic rays.

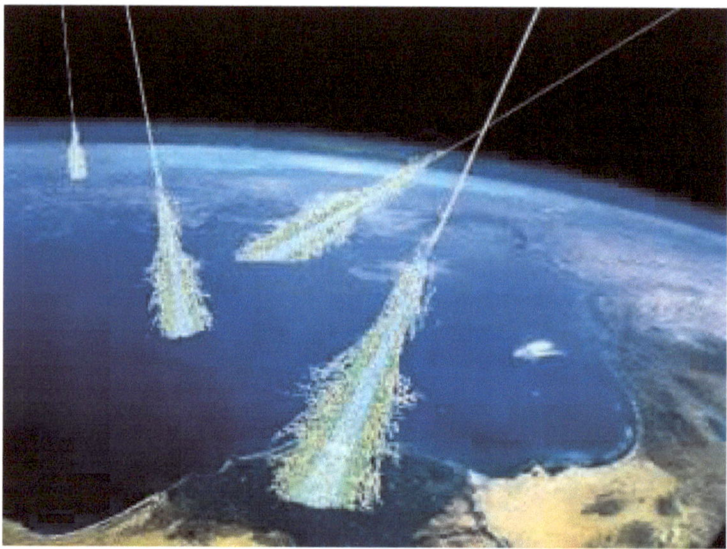

Fig. (9.1). Air showers from very high-energy cosmic rays on Earth's atmosphere. (Image credit: Simon Swordy (U. Chicago), NASA) Source:https://cdn.mos.cms.futurecdn.net/JZG8KC9eZovtTaXkvbbdiX-97--80.jpg.webp

Since cosmic rays convey electric charge, their direction deviates as they pass through magnetic fields. By the time the particles reach us, their paths are completely jumbled, as shown in fig. (**9.2**) by the blue path. We can't trace them back to their sources. Light journeys to us straight from their bases, as shown in fig. (**9.2**) by the purple path.

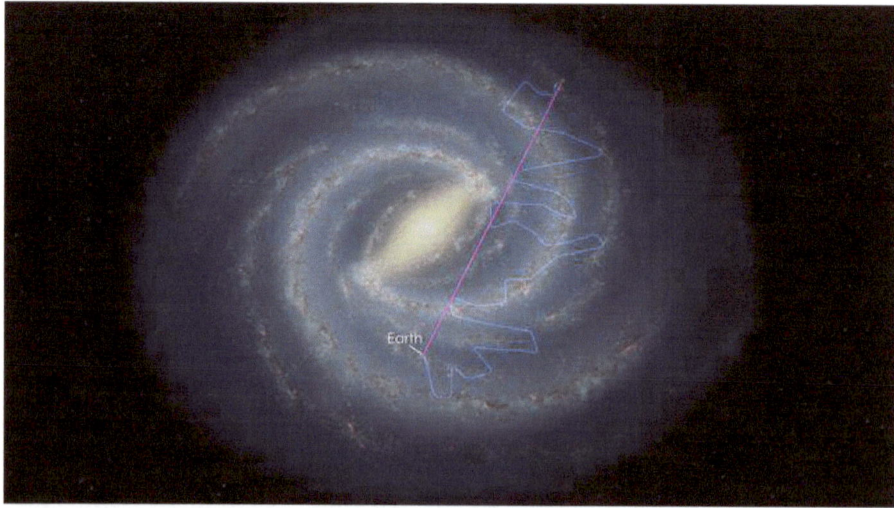

Fig. (9.2). Path of cosmic rays and light. (Image Credit: NASA's Goddard Space Flight Center). Source: https://imagine.gsfc.nasa.gov/Images/science/cosmic_ray_path.png.

Hence understanding the source of cosmic rays using direct measurements is a difficult assignment due to the existence of intergalactic magnetic fields that diverge the path of these stimulating particles before they touch Earth's surface. Impersonal couriers produced by cosmic-ray connections, such as neutrinos and gamma rays, are used to control the source of cosmic rays (Fig **9.3**).

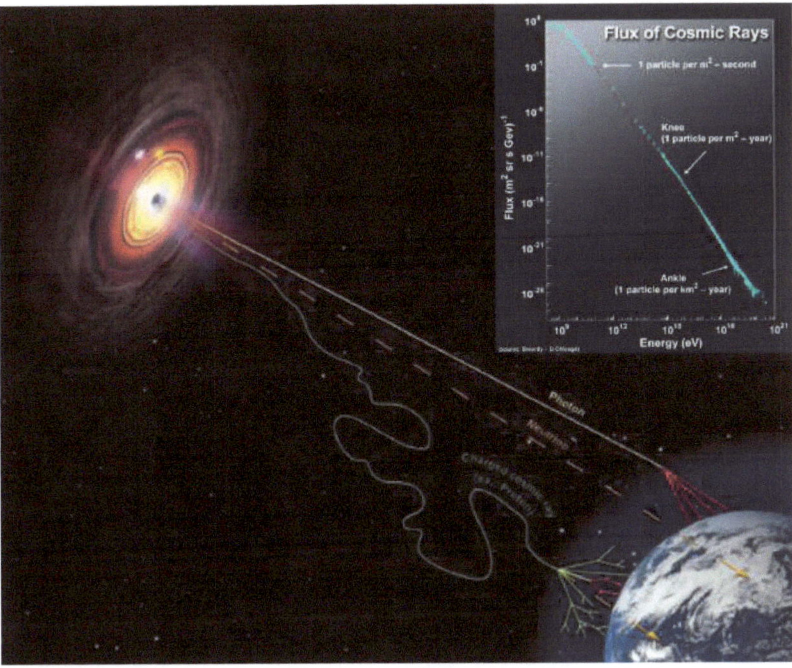

Fig. (9.3). The origin of cosmic rays is discovered with neutral couriers, such as gamma rays. The enclosed figure on the top right shows the cosmic-ray energy spectrum perceived from Earth with the foremost spectral features. (Image Credit: HAP/A. Chantelauze). Source:https://www.cta-observatory.org/wp content/uploads/2020/08/ScienceArticle_WH_RO_AFB.docx-1536x1278.png.

1.1.1. Primary Cosmic Rays

Primary cosmic rays (CR) can be accelerated by astrophysical sources like Protons ~87%,He ~12% and 1% of heavier nuclei such as Fe, O, C, supplementary ionized nuclei produced in stars, 2% electrons γ rays, neutrinos. Moreover, there may be primary components of anti-p and e+ (antimatter in the Universe?) But composition varies with energy (the bulk of CR is at 1 GeV). Primary cosmic rays mostly originate from outside the Solar System and sometimes even the Milky Way. When they interact with Earth's atmosphere, they are adapted to secondary particles. The mass ratio of helium to hydrogen nuclei, 28%, is comparable to the elemental abundance [24%] ratio of these components [12]. The residual portion is made up of the other denser nuclei that are characteristic nucleosynthesis end products, predominantly beryllium, lithium and boron. These nuclei look as if cosmic rays are in much superior abundance (≈1%) than in the solar sky, where they are only approximately 10−11 as plentiful as helium. Cosmic rays made up of charged nuclei heavier than helium are called HZE ions. Due to the high charge and heavy nature of HZE ions, their contribution to a spaceman's radiation amount in space is substantial; however, they are moderately rare.

This abundance transformation is a consequence of the way secondary cosmic rays are made. Oxygen and carbon nuclei strike with intergalactic matter to form beryllium, lithium and boron in development characterized by cosmic ray spallation. Spallation is also accountable for the profusions of titanium, vanadium, scandium, and manganese ions in cosmic rays formed by explosions of nickel and iron nuclei with astronomical matter [13]. At high energies, the configuration deviates, and denser nuclei have higher richness in some energy assortments. The existing investigation goal is focused on more precise measurements of the composition at high energies. They are a flux of stable (>106 years) charged particles and nuclei. Fig. (**9.4**) indicates the primary cosmic ray spectrum as a function of the energy.

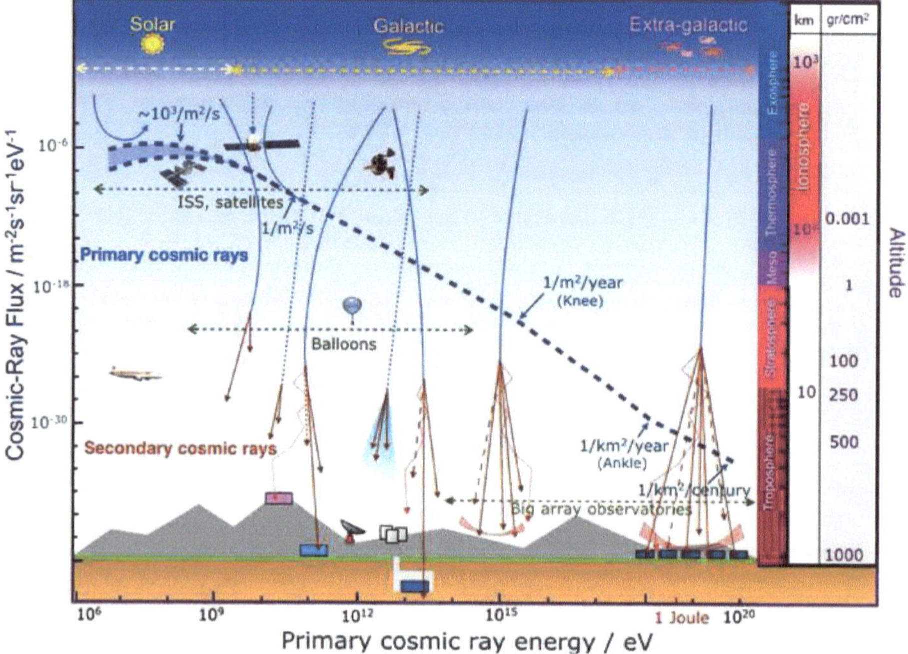

Fig. (9.4). Primary cosmic ray spectrum as a function of the energy. Image Credit: Ref [14]: Source: https://www.researchgate.net/profile/Anna-Morozova4/publication/312914838/figure/fig6.AS:668318931353604@1536351058355/Primary-cosmic-ray-spectrum-as-a-function-of-the-energy-The-blue-dashed-line-represents.jpg.

1.1.2. Secondary Cosmic Rays

These are the particles formed in exchanges of primary cosmic rays with intergalactic gas and also constituent fragments (particles) shaped in the sky (Be, B, Li, e^+, anti-p). Separately from elements created in solar flares, they come from outside the Solar System. When cosmic rays arrive on the Earth's atmosphere, they strike atoms and molecules, primarily oxygen and nitrogen. The collaboration yields a flow of lighter particles, a so-called air shower secondary radiation that showers down, including protons, neutrons, electrons, pions, muons, x-rays, alpha particles and neutrinos [15]. Fig. (**9.5**) indicates schematics of gathering secondary cosmic rays in the atmosphere.

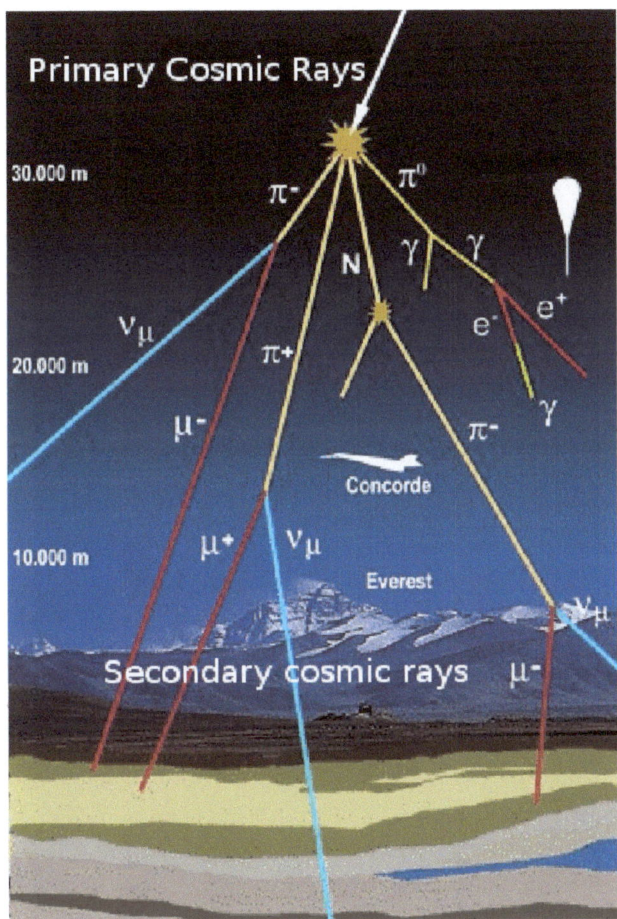

Fig. (9.5). Gathering of Secondary Cosmic Rays in the Atmosphere. Image Credit: Ref [14]. Source: https://www.researchgate.net/profile/Francisco-Barradas-Solas/publication/264310353/figure/fig2/AS:667619187240963@1536184226142/Production-of-Secondary-Cosmic-Rays-in-the-Atmosphere_Q640.jpg

All of the secondary particles formed by the smash remain forward on paths inside almost one degree of the primary particle's initial path. Characteristic particles created in such impacts are neutrons and stimulating mesons such as positive or negative pions and kaons. Some of these consequently fall off into muons and neutrinos, capable of touching the surface of the Earth. Some high-energy muons even enter for some distance into reedy excavations, and most neutrinos pass through the Earth without additional interaction. Others degenerate into photons, consequently generating electromagnetic flows. Hence, resulting in photons, positrons and electrons typically govern air showers. These particles and muons can be easily identified by several types of particle detectors, such as scintillation or water-Cherenkov detectors, bubble chambers, and cloud chambers. The observation of a secondary shower of particles in numerous detectors at the same time is a symptom that all of the particles came from that incident. Cosmic rays controlling other terrestrial bodies in the Solar System are identified secondarily by detecting high-energy gamma-ray discharges by gamma-ray telescope. These are well-known for radioactive decay developments by their higher energies above about 10 MeV [15].

1.2. Composition of CRs in the Solar System and the Galaxy

All stable elements of the periodic table are found in large quantities of CRs and are identically analogous to the solar structure. Convincing with a great quantity of Silicon as reference (by characterization, its profusion is expected equivalent for both - it is stress-free to computation), the comparative resources of the components in the solar coordination and enormous CRs are linked, and it is perceived that:

i. CRs comprise a smaller amount of He and H than in the solar system
ii. CRs encompass more light elements (B, Li, Be) than a solar system
iii. In CRs, the extent of unusual (odd) Z components is larger (odd-even effect)
iv. In CRs, there are additional sub-Fe components

Be, B, and Li are not created in star nucleosynthesis but are the outcome of spallation on an astronomical matter identical to collisions between IM and CRs. Fig. (**9.6**) indicated an abundance of elements ($Z<28$) in CRs[43,36] at 1 TeV. These are split individual follow-on nuclei with charge and mass numbers just less than those of the mutual components like H and He [16]:

Subsequently, for H and He, Atomic No. Z = 1, it is challenging to ionize and speed up them. Table **9.1** reveals the relative and absolute abundance of particles in cosmic rays and the Universe.

Table 9.1. Relative and absolute abundance of particles in Cosmic Rays and The Universe.

Particle Group	Nucleus Charge	Integrated Intensity of Particle ($m^{-2}\,s^{-1}\,sr^{-1}$)	Number of Particles per 10^5 Protons	
			In Cosmic Rays	In the Universe
protons	1	1300	10000	10000
helium	2	94	720	1600
L	3-5	2	15	10^{-4}
M	6-9	6.7	52	14
H	10-19	2	15	6
VH	20-30	0.5	4	0.06
SH	> 30	10^{-4}	10^{-3}	$7 \cdot 10^{-5}$
electrons	1	13	100	10000
antiprotons	1	> 0.1	5	???

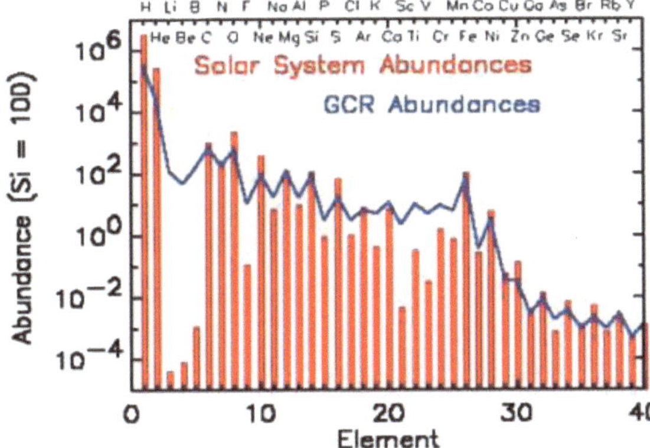

Fig. (9.6). Abundance of elements (Z<28) in CRs[43,36] at 1 TeV.

The odd/even influence is because nuclei with odd Z and/or A are meager bound and more recurrent produces in thermonuclear responses. Particularly steady nuclei arise for filled shells (‘magic nuclei’) analogous to magic numbers (2, 8,20,50,82,126) that mention independently to n and p. Paired (twofold) magic nuclei like He and O are predominantly steady and hence rich.

i. The improved wealth of B, Be, and Li in CRs is due to the spallation of denser constituents (such as C and O). Sub-Fe originated from the breaking up of Fe, which is moderately rich.
ii. Spallation exchanges and subsequent richness give astronomical depth and a typical CR lifetime.

Although not shown in the figure, many elements heavier than the iron group have been dignified with characteristic abundances of 10−5 of iron. Much of this evidence was gained from satellite and spacecraft measurements over the last decade. Some of the conclusions:

1. Abundances of even Z elements with $30 \lesssim Z \lesssim 60$ are in practical agreement with solar system abundances.
2. In the region $62 \lesssim Z \lesssim 80$, which comprises the platinum-lead region, abundances are improved compared to solar by about an aspect of two. This recommends an improvement in r-process elements, which govern this mass region. There are obvious influences amongst other astrophysics we have discussed earlier and possible deviations from solar abundances in the cosmic rays [17].

2. ELEMENTARY PARTICLES

Elementary particles are elements with no anticipated inner configuration; that is, it is unidentified whether they are composed of other elements [18]. They are the essential entities of quantum field theory. Many families and sub-families of elementary particles be existent. Elementary particles are categorized authorizing to their spin. Fermions have half-integer spin while bosons have integer spin. The whole thing related to particles of the Standard Model has been experimentally witnessed in recent times containing the Higgs boson in 2012 [19]. Many other theoretical elementary particles, such as the graviton, have been projected but not perceived experimentally [20].

About 1980, an elementary particle's prominence was certainly fundamental as a final integral of substance and was ordinarily rejected for a more useful outlook [18] personified in particle physics' Standard Model, what's known as science's most experimentally effective theory [21]. Many explanations and theories out there; the Standard Model, containing the popular supersymmetry, double the number of elementary particles by assuming that each known particle links with a "shadow" partner far more enormous [22, 23], even though all such superpartners stayed unexposed [24]. In the meantime, an elementary boson predominates gravitation, so the graviton remains theoretical. Also, as hypotheses indicate, space-time is perhaps quantized, so there most likely exist "atoms" of space and time themselves [25].

Fig. (**9.7**) reveals the Standard Model of elementary particles, including the 12 fundamental fermions and 5 fundamental bosons, including Higgs bosons.

2.1. Leptons and Quarks

They are fundamental and sub-atomic particles incapable of being subdivided into smaller particles.

i. There are 6 Leptons and 6 Quarks.
ii. The nucleus is an assembly of quarks that notices themselves as protons and neutrons.
iii. Each elementary particle has a corresponding antiparticle.

2.1.1. Leptons

Leptons and quarks are the basic building blocks of matter, *i.e.*, they are understood as the "elementary particles." There are six leptons in the contemporary structure, the electron, muon, and tau particles and their accompanying neutrinos. The collections of the elementary particles are characteristically called "flavors," and the neutrinos here are considered to have remarkably unlike flavors.

ImageSource:https://i.pinimg.com/originals/50/ce/97/50ce976f1415d03361ed98a99ae08a2d.png

In particle physics, a lepton is an elementary particle of half-integer spin (spin 1⁄2) that does not experience strong interactions. Two core constituents of leptons exist charged leptons (also acknowledged as the muons or electron-like leptons) and neutral leptons (identified as neutrinos). Charged leptons can combine with supplementary particles to make a system of various composite particles such as atoms and positronium, whereas neutrinos barely work collectively with anything and are thus rarely observed. The best acknowledged of entirely discovered leptons is the electron. Leptons have numerous inherent properties, comprising mass, electric charge and spin. Unlike quarks, on the other hand, leptons are not exposed to the strong interaction, but they are exposed to the other three fundamental dealings: gravitation, the weak interaction, and electromagnetism, of which the latter is comparative to charge and is thus zero for the electrically unbiased neutrinos. For every lepton flavor, there is a consistent type of antiparticle, known as an antilepton, which varies from the lepton in that some of its possessions have equivalent magnitude but reverses signs. According to certain theories, neutrinos may be their specific antiparticle [26]. Leptons are a significant fragment of the Standard Model. Electrons are one of the constituents of atoms, together with protons and neutrons. Unusual atoms with muons and taus as an alternative to electrons can also be produced, as well as lepton–antilepton particles such as positronium [27].

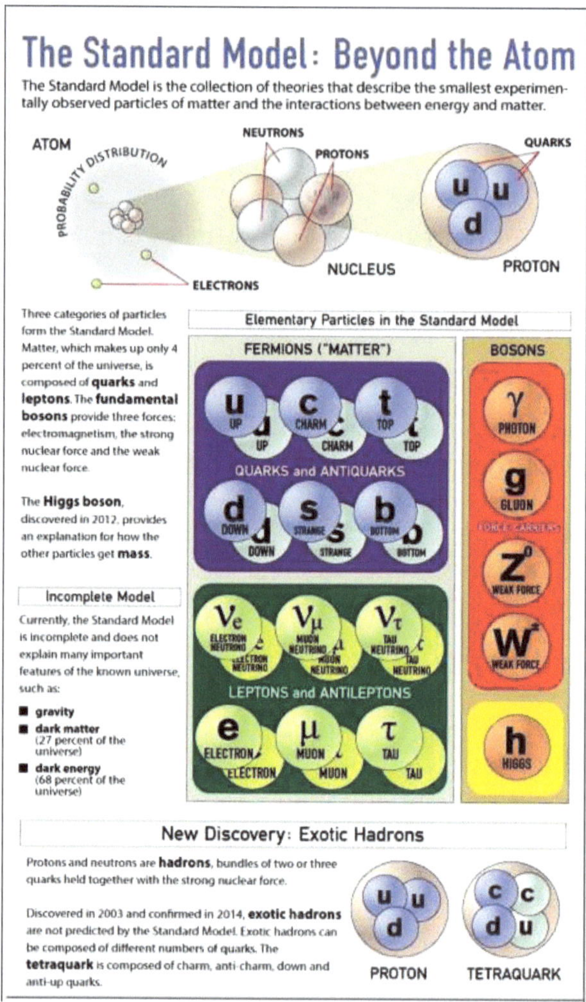

Fig. (9.7). The Standard Model of elementary particles. Image Credit: Karl Tate, Livescience.com Infographics Artist). Source: https://cdn.mos.cms.futurecdn.net/rC4eKu5DhiDDMkXAH85FJV-970-80.jpg.webp.

The positron is the antiparticle of the electron, and when a positron enters any normal matter, it will find an abundant amount of electrons with which to annihilate. The energy released by the annihilation produces two highly energetic gamma rays, as shown in Figs. (**9.8** and **9.9**), and if one assumes that the momenta of the positron and electron were identical before the annihilation, the two gamma ray photons must travel in opposite directions in order to conserve momentum.

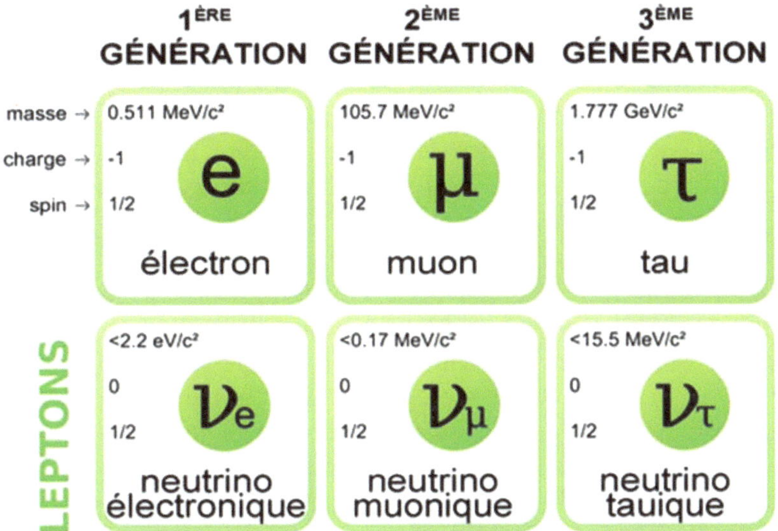

Fig. (9.8). Schematics of Lepton flavors.

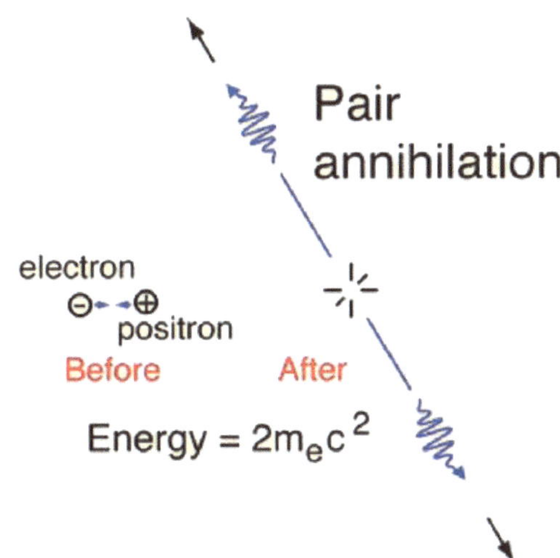

Fig. (9.9). Annihilation and production of two highly energetic gamma rays.

2.1.2. Quarks

The Standard Model is the hypothetical outline unfolding all the identified elementary particles. Present model encompasses six flavors of quarks (q), named as, up (u), down (d), strange (s), charm (c), bottom (b), and top (t) [28]. Antiparticles of quarks are entitled antiquarks and are signified by a bar over the symbol for the consistent quark, such as u for an up antiquark. As with antimatter in broad-spectrum, antiquarks have the identical mass, mean lifetime, and spin as their corresponding quarks, but the electric charge and other charges have the opposite sign [29]. Quarks are spin-1/2 particles, inferring that they are fermions authorizing the spin-statistics theorem. They are subjected to the Pauli Exclusion Principle, which statuses that no two identical fermions can simultaneously occupy the same quantum state. This is in divergence to bosons (particles with integer spin), of which any number can be in the same state [30]. Unlike leptons, quarks have a color charge, which sources them involved in the strong interaction. The subsequent attraction amongst different quarks origins the development of complex particles well-known as hadrons (see "Strong interaction and color charge" below). Electric charge, mass, color charge, and flavor quarks are the only recognized elementary particles that are involved in the entire four fundamental interactions of current physics: electromagnetism, gravitation, strong interaction, and weak interaction [31]. Gravitation is weak to be appropriate to distinct particle networks distant from Planck energy (extremes of energy) and Planck distance. Nevertheless, no productive quantum theory of gravity exists in existent physics; gravitation is not pronounced by the Standard Model. Fig. (**9.10**) displays schematics of quark masses for all six flavors in an association.

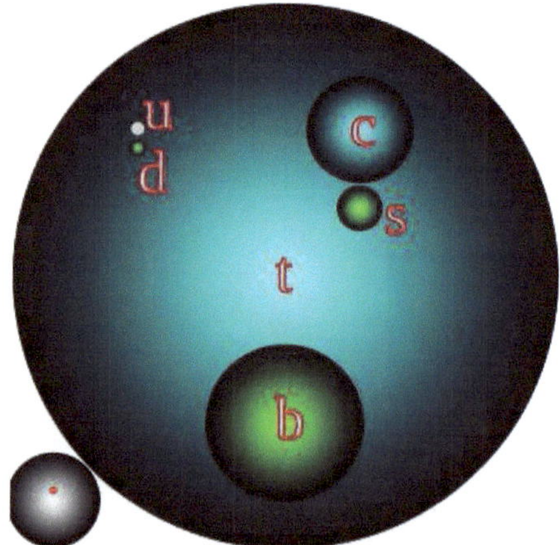

Fig. (9.10). Current quark masses for all six flavors in comparison, like balls of proportional volumes. Proton (gray) and electron (red) are shown in the bottom left corner for scale. Image Credit: Incnis Mrsi, CC BY-SA 3.0 <https://creativecommons.org/licenses/by-sa/3.0>, *via* Wikimedia Commons. Source: https://upload.wikimedia.org/wikipedia/commons/b/b5/Quark_masses_as_balls.svg.

2.2. Matter and Antimatter

Antimatter is identical to normal matter in almost every way. The only difference is electric charge, which is the opposite of the two forms of matter. So there could be a whole galaxy made of antimatter out there, and our telescopes wouldn't see it any differently from a galaxy of normal matter.

Most theories say the Big Bang should have created equal amounts of matter and antimatter. But something transformed that balance in the first tiny fraction of a second. For every billion pairs of matter and antimatter particles, there was one extra particle of matter. The difference between these two forms of matter is more fundamental than it appears. What we call matter is everything that is composed of protons (a sub-atomic particle with a positive charge), electrons (a sub-atomic particle with a negative charge), and neutrons (a sub-atomic particle with no charge). All these particle systems are what we call atoms. In the atom, the protons and neutrons make up the nucleus, which is the core, and the electrons orbit the nucleus much like a planet around a star.

In antimatter, the charges of each particle are reversed. Instead of a proton, its antimatter equivalent is called an anti-proton with a negative charge. Instead of an electron, its antimatter equivalent is called a positron with a positive charge. The exception to this rule is the neutron, whose antimatter counterpart, the anti-neutron, shares the same traits (since a neutron has no charge, its anti-form would retain no charge) [32].

If one were to combine antimatter and matter, it would generate an enormous explosion of energy. This is produced by linking the opposite charges of each complement, which thus origins them to be overturned into the form of energy based on the equation $E=mc^2$, e meaning energy, m equaling mass, and c equals the speed of light, approximately 186,000 miles per second. But not to worry, since the only method of generating antimatter on Earth involving particle accelerators only yields a few particles at a time, thus avoiding any unsuccessful reactions.

In broad-spectrum chemistry and conventional physics, the material is any matter that has mass and occupies the cosmos (space) by having volume [33]. Entire everyday entities that can be touched are ultimately composed of atoms, which are made up of subatomic particles that work collectively, and in scientific as well as everyday practice, "matter" commonly embraces atoms and whatsoever made up of them, and any particles (or grouping of particles) that act as if they have both rest mass and volume. On the other hand, it does not contain massless particles such as photons or waves such as light or other energy occurrences [34]. Matter occurs in numerous states (also acknowledged as phases). These include established classical and routine phases such as solid, liquid, and gas – for example, water may be present as ice, liquid water, and gaseous steam – but supplementary states are also probable, including plasma, quark–gluon plasma, fermionic condensates and Bose–Einstein condensates [35]. Fig (**9.11**) discloses a composite astronomical image of antimatter being left behind after a collision of galaxies (Cluster Abell 520).

Fig. (9.11). Composite astronomical image of dark matter being left behind after a collision of galaxies (Cluster Abell 520). Public Domain Image, source: NASA, ESA, CFHT, CXO, M.J. Jee (University of California, Davis), and A. Mahdavi (San Francisco State University). Image Source: https://wtamu.edu/~cbaird/sq/images/hs-2012-10-a-large_web.jpg

In Fig. (**9.11**), the starlight is colored white and orange, showing the location of the galaxies. The hot gas from the collision is colored green. The dark matter left behind is colored blue and is determined by gravitational lensing measurements.

Antimatter particles mix up with each other to form antimatter; just as usual, particles bind to form normal matter. For example, a positron (the antiparticle of the electron) and an antiproton (the antiparticle of the proton) can form an antihydrogen atom. The nuclei of antihelium have been produced beforehand, even with great concern, and are the utmost complex anti-nuclei so far tried [36]. Physical principles designate that complex antimatter atomic nuclei are imaginable and anti-atoms reliable to the known chemical elements. There is a healthy suggestion that the apparent cosmos is self-possessed and essentially made up of entire ordinary matter complementing an indistinguishable combination of matter and antimatter. This irregularity of matter and antimatter in the observable cosmos is one of the great unexplained problems in physics. The procedure by which this distinction between matter and antimatter particles is recognized is called baryogenesis [37]. Table **9.2** indicates.

Table 9.2. The comparison between matter and antimatter.

Regular Matter	Antimatter
Examples: electron, proton, neutron Main Role: forms atoms, molecules, objects, planets, *etc.* Reflects Light?: yes	Examples: positron, antiproton Main Role: annihilates regular matter Reflects Light?: yes

2.3. Characteristics of Elementary Particles

2.3.1. Leptons

Leptons are elementary particles with spin 1/2 (a fermion) that are not exaggerated by the strong nuclear force. They are an assembly of particles that are improved from the other known assembly of fermions, called the 'quarks.'

Electrons are a renowned example that is found in conventional matter. There are six leptons: the electron, muon, and tau particles and their concomitant neutrinos. The different selections of the elementary particles are called "flavors," and the neutrinos here are deliberated to have a particularly altered flavor. Three of the six leptons have an electric charge, and three do not. The best acknowledged charged lepton is the electron (e). The other two charged leptons are the muon (μ) and the tau (τ), which are like electrons but considerably larger. The charged leptons are all negative particles; their antiparticles are positively charged (for example, the antiparticle of the electron, e-, is a positron, e+). Moreover, they are elementary particles and do not have measurable size or structure. The neutrinos do not have an electric charge, and each of the six has an anti-particle.

2.4. Electron, Muon and Tau

2.4.1. Electron

The electron is a subatomic particle, symbol $e-$ or $\beta-$. Electrons have their place in the first group (generation) of the lepton particle intimate and are in general made up to be elementary particles for the purpose that they have no recognized substructure or elements. The electron has a mass that is nearly 1/1836 that of the proton. Quantum mechanical possessions of the electron consist of a spin (an intrinsic angular momentum) of a half-integer value (1/2), stated in units of the condensed Planck constant, ħ. Being fermions, permitting to the Pauli exclusion principle, no two electrons can occupy the same quantum state. Approximating all elementary particles, electrons signify properties of both particles and waves: they can strike with other particles and can be repelled similar to photons or light. The wave properties of electrons are stress-free to detect with experimentations than those of other particles like neutrons and protons since electrons have a lower mass and an extensive de Broglie wavelength for particular energy [38].

2.4.2. Muon

The muon is an elementary particle. It has a negative electric charge and half spin (1/2). Its symbol is "μ-." It is a type of lepton. The muon has a normal lifetime of 2.2 micro-seconds or 0.0000022 seconds. A muon is comparable to an electron, but it is 200 times more compressed and heavier. Unlike an electron, it decays into other particles. The superparticle of a muon is called a "smuon."

As with the decay of the non-elementary neutron with a lifetime of about 15 minutes, by subatomic standards, muon decay is undisturbed since the decay is settled only by the weak interaction and since the mass difference between the muon and the set of its decay products is less significant, provided that they resemble limited kinetic degrees of freedom for decay. Muon decay practically yields at least three particles, which essentially encompass an electron of the equivalent charge as the muon and two types of neutrinos.

Like all elementary particles, the muon has a corresponding antiparticle of opposite charge (+1 e) but equal mass and spin: a positive muon is known as the antimuon. Muons are signified by μ− and antimuons by μ+.Previously, muons were termed "mu mesons" but are not categorized as mesons by contemporary particle physicists, and that name is no longer used by the physics community. Muons have a mass of 105.66 MeV/c^2, which is approximately 207 times that of the electron [39].

Due to their larger mass, muons accelerate more progressively than electrons in electromagnetic fields, emitting a smaller amount slowing down (attenuating) radiation. This permits muons of a given energy to enter far deeper into the matter since the attenuation of electrons and muons is mainly due to energy loss by the bremsstrahlung mechanism (or braking radiation produced due to the deceleration (negative acceleration) of a charged particle). For example, pretended "secondary muons," produced by cosmic rays striking the atmosphere, can enter the atmosphere and spread to Earth's land surface and even into deep mines.

2.4.3. Tau

The tau (τ), also called the tau lepton, tau particle, or tauon, is an elementary particle similar to the electron, with a negative electric charge and a spin of 1/2. Like the electron, the muon, and the three neutrinos, the tau is a lepton, and like all elementary particles with half-integer spin (1/2), the tau has an analogous antiparticle of opposite charge but equal mass and spin. In the tau's example, this is the positive tau, also called the "antitau." Tau particles are denoted by the symbol τ− and the antitaus by τ+. Tau leptons have a lifespan of 2.9×10^{-13} seconds and a mass of 1776.86 MeV/c^2 (matched to 105.66 MeV/c^2 for muons and 0.511 MeV/^{c2} for electrons). Subsequently, their exchanges are comparable to those of the electron; a tau can be believed to be a much heavier type of the electron. Since of their larger mass, tau particles do not discharge as much bremsstrahlung radiation as electrons; accordingly, they are theoretically much stronger than electrons. Because of its short lifetime, the range of the tau is primarily set by its decay length, which is excessively lesser for bremsstrahlung to be visible. Its probing power performs only at ultra-high velocity and energy; when time expansion prolongs, it then has a very short path length. As with the case of the other charged leptons, the tau has an associated tau neutrino, denoted by ντ.

In short, we can associate Electron, Muon and Tau as:

- All three have a charge of -1
- The electron is found in everyday matter
- The muon and the tau have a lot more mass than the electron
- The muon and the tau are not part of everyday matter because they have very short lifetimes

Figs. (**9.12 a,b and c**) indicates the schematics of Electron, Muon and Tau, respectively.

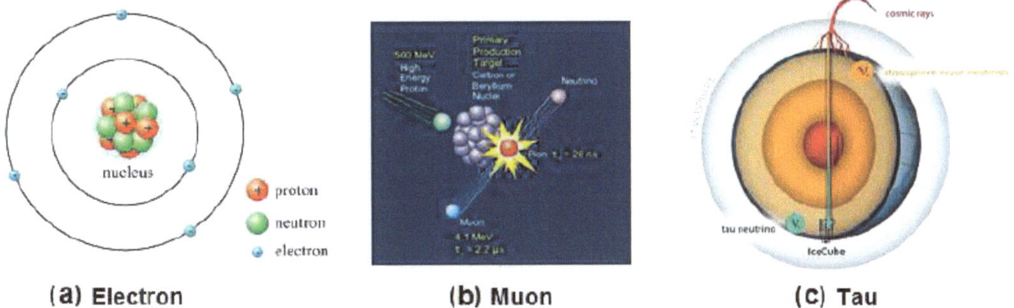

Fig. (9.12). Schematics of Electron, Muon and Tau.

a. Schematics of an Electron in Atom

Image Credit: Healing Tones, The atom Image source:
https://attunementwithsacredsound.files.wordpress.com/2018/02/ the-atom.gif

b. An Illustration of How Muons Are Formed

Image Credit: Dylon Martin, Dylon's Corner: The Wonderful World of μS, TRIUMF: Canada's particle accelerator centre.

Image Source: https://www.triumf.ca/sites/default/files/styles/img_assist_style/public/images/Week2-figure1.JPG?itok=F0YjG8ru

c. Atmospheric Tau Neutrino Appearance

Image Credit: Ice Cube Neutrino Observatory - UW-Madison Image Source: https://res.cloudinary.com/icecube/images/q_auto/w_1110,h_883/v1608000259/news_attachment.file_.a1db53fbc7338fb9.4d756f6e746f5461755f53747261696768742e6a7067/news_attachment.file_.a1db53fbc7338fb9.4d756f6e746f5461755f53747261696768742e6a7067.jpg

2.4.4. Neutrinos

Neutrinos are a type of elementary particle that exist all across the cosmos. The word neutrino means a minor neutral particle. Physicists study these particles, but they are challenging to find since they have a very small chance of interacting with steady matter. (For example, they typically pass through the entire globe without touching any other particles). Neutrinos travel near the speed of light. Electrons have an electric charge, but neutrinos do not. This means that neutrinos are unaffected by the electromagnetic force. The first detectors were built to find neutrinos experientially, only 10 or 15 each year. Modern finders use chambers that comprise around a thousand tons of water or other liquid, which lets them identify about 10 to 100 neutrinos per day [40]. Neutrinos are produced in particle accelerators, in the sun, in other stars, and in nuclear reactions such as in nuclear reactors. They are produced when there is a nuclear reaction in the form of beta decay. This process starts off with one neutron and ends with one electron, one proton, and one neutrino.

- Neutrinos are three of the six leptons
- They have no electrical or strong charge
- Neutrinos are very stable and are all around
- Most neutrinos never interact with any matter on Earth

Fig (**9.12b**) exposes the initial use of a hydrogen bubble chamber to recognize neutrinos, on 13 November 1970, at Argonne National Laboratory. At this juncture, neutrino knocks out a proton in a hydrogen atom; the impact takes place at the point where three pathways originate on the right of the photograph.

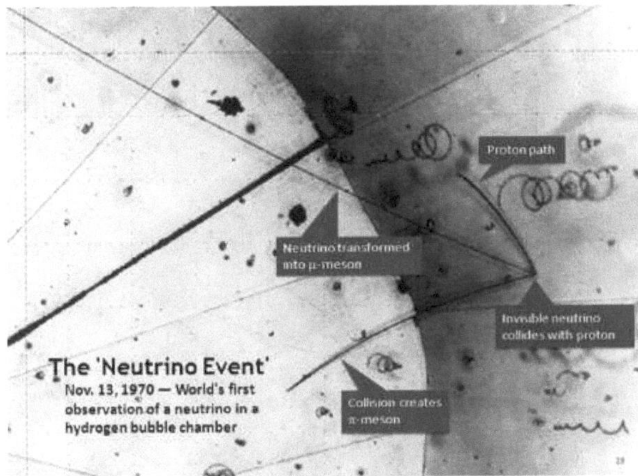

Fig. (9.12b). The first use of a hydrogen bubble chamber to detect neutrinos. Image Source: https://upload.wikimedia.org/wikipedia/commons/5/57/FirstNeutrinoEventAnnotated.jpg.

2.4.5. Quarks

Lonely quarks and antiquarks have never been noticed, information clarified by confinement. Every quark transports one of three color charges of the strong interaction; antiquarks similarly carry anticolor. Color-charged particles act together *via* gluon conversation in the same way charged particles act together *via* photon conversation. Nevertheless, gluons are themselves color-charged, ensuing in an intensification of the strong force as color-charged particles are disconnected. Unlike the electromagnetic force, which reduces as charged particles are isolated, color-charged particles feel a growing force.

Quarks also carry insignificant electric charges, but, subsequently, they are restricted within hadrons whose charges are all integral; fractional charges have never been quarantined. Note that quarks have electric charges of either $+2/3$ or $-1/3$, while antiquarks have equivalent electric charges of either $-2/3$ or $+1/3$. Confirmation for the presence of quarks originates from deep inelastic scattering: firing electrons at nuclei to regulate the scattering of charge within nucleons (baryons). If the charge is identical, the electric field nearby the proton should be constant, and the electron should scatter elastically. Low-energy electrons scatter this way, but above precise energy, the protons bounce some electrons through enormous directions. The jump-back electron has much less energy, and a jet of particles is radiated. This unbendable scattering recommends that the charge in the proton is not identical but fragmented among smaller charged particles, well known as quarks. The fundamental quark retains the following properties.

- Quarks are elementary particles and are used to create other particles
- Each quark has an anti-particle

- Quarks have a physical property called color; they could be blue, green or red
- Each color also has an anti-color
- They are not really different colors; it is a property, like a charge
- Quarks cannot exist individually because the color force increases as they are pulled apart.
- The important six quarks are: Up, Down, Strange, Charm, Bottom and Top

Fig. (**9.13**) reveals the schematics for the generation of quarks form the matter, molecule and atom.

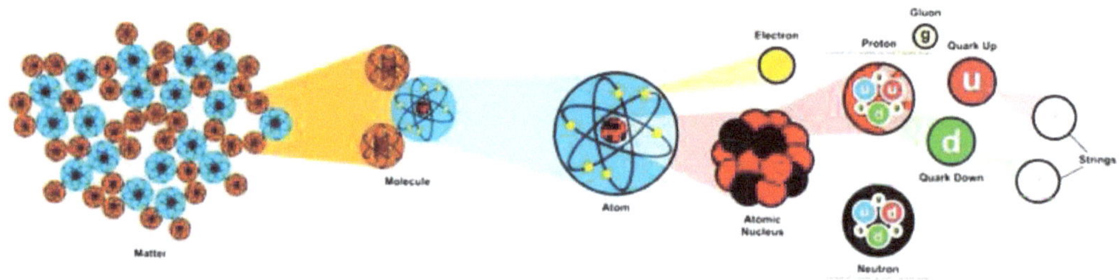

Fig. (9.13). Infographic diagram showing the smallest particle (Quark) discovered so far. Image Source: https://image.shutterstock.com/image-vector/form-matter-molecule-atom-quarks-600w-682932847.jpg.

2.4.6. Hadrons

In particle physics, a hadron is a subatomic complex particle prepared of two or more quarks held together by the strong force in a comparable way as molecules are held together by the electromagnetic force. Most of the mass of regular matter originates from two hadrons: the proton and the neutron. Hadrons are categorized into two families: baryons, made of an odd number of quarks, usually, three quarks and mesons, made of an even number of quarks, usually one quark and one antiquark [41]. Fig. (**9.14**) reveals schematics of Hadrons.

The bulk mass of an atom is made up of protons, and neutrons are illustrations of baryons; pions fall under the pattern of a meson. "Mysterious" hadrons, surrounding more than three valence quarks, have been wide-opened in recent years. A tetraquark state (an exotic meson), named the Z (4430), was uncovered in 2007 by the Belle Collaboration [42] and recognized as a consequence in 2014 by the LHC teamwork [43]. There are plenty more interesting hadron participators and other color-singlet quark alliances that may also exist.

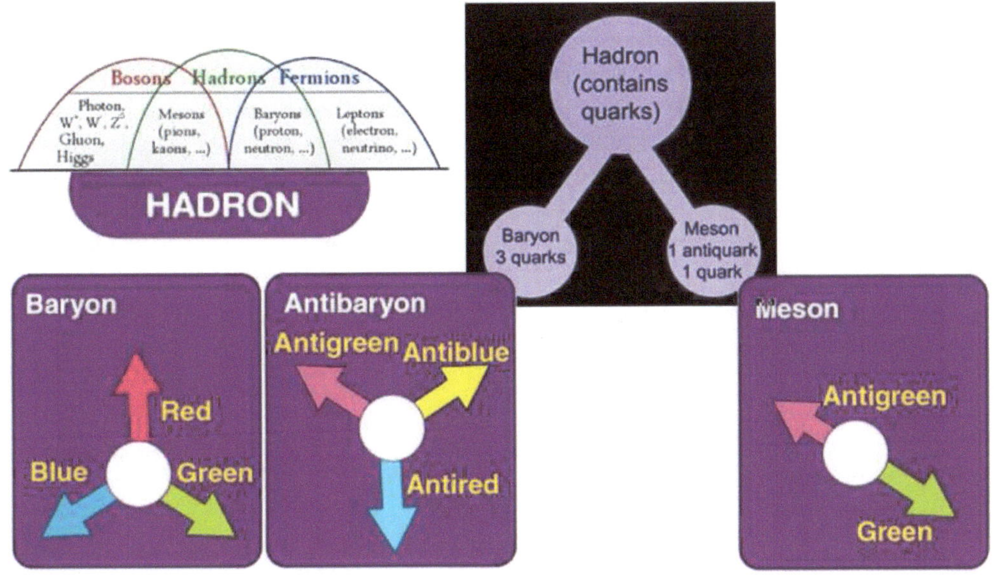

Fig. (9.14). Schematics of Hadrons.

Practically altogether, "free" hadrons and antihadrons are thought to be unstable and eventually break down into other constituent parts. The only predictable authorization interconnects to free protons, which are conceivably steady, or at minimum, take massive groupings of time to degeneration. Free neutrons are unbalanced and degenerate with a half-life of about 611 seconds. Their corresponding antiparticles are predictable to trail the identical arrangement, but they are challenging to arrest and study since they instantly suppress interaction with conventional matter. "Bound" protons and neutrons, confined surrounded by the atomic nucleus, are usually considered steady. Experimentally, hadron physics is deliberated by striking protons or nuclei of heavy elements such as lead or gold and identifying the wreckage in the created particle showers. In the normal atmosphere, mesons such as pions are formed by the impacts of cosmic rays on the atmosphere. These are distinguished by their internal structure. Most of the mass we observe in a hadron comes from its kinetic and potential energy.

2.4.7. Baryons

A baryon is a category of complex subatomic particle which encompasses an odd number or at least 3 valence quarks [44]. Baryons have their place in the hadron family of particles; hadrons are a collection of quarks. Baryons are also categorized as fermions since they have half-integer spin. The name "baryon," announced by Abraham Pais [45], originates from the Greek word for "heavy" for the reason that, at the time of their identification, most recognized elementary particles had lower masses than the baryons. Each baryon has an equivalent antiparticle (antibaryon) where their analogous antiquarks substitute quarks. For example, a proton is made of two up quarks and one down quark; and its analogous antiparticle, the antiproton, is made of two up antiquarks and one down antiquark. Since they are composed of quarks, baryons contribute to the strong interaction, which is intermediated by particles well-known as gluons. The most familiar baryons are protons and neutrons, both of which encompass three quarks, and for this reason, they are occasionally called triquarks. These particles make up most of the mass of the observable matter in the universe and compose the nucleus of every atom.

- Baryons are composed of three quarks
- All but two baryons are very unstable; they are: The proton and The neutron
- Most baryons are excited states of protons and neutrons
- Protons are made of three quarks, two up quarks and a down quark, and this is written as 'uud'. However, neutrons are also made up of three quarks, one up quark and two down quarks and this is written as 'udd' as shown in Fig. (**9.15**).

(The excited states of protons and neutrons)

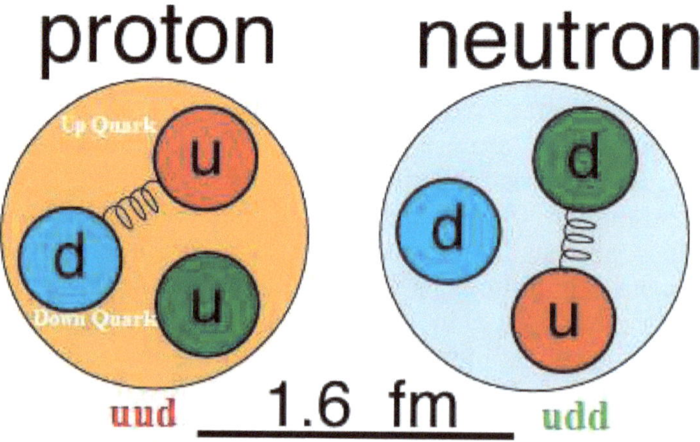

Fig. (9.15). Schematics illustration of Baryons.

2.4.8. Mesons

In particle physics, mesons are subatomic particles originating from hadrons made up of an equal number of quarks and antiquarks, typically one of each, connected by strong interactions. Since mesons are composed of quark sub-particles, they have a significant physical size, a diameter of approximately one femtometer

(1×10−15 m) [46] which is around 0.6 times the size of a proton or neutron. All mesons are unsteady, with the longest-lived persistent for merely a few hundredths of a microsecond. Denser mesons degenerate to lighter mesons and eventually to steady electrons, neutrinos and photons. Outside the nucleus, mesons appear in natural surroundings only as short-lived produces of very high-energy collisions amongst particles made of quarks, high-energy protons and neutrons, baryonic matter and cosmic rays. Mesons are characteristically created artificially in cyclotrons or other accelerators in the crashes of protons, antiprotons, or other particles.

Other massive or higher-energy mesons were produced for a short time in the Big Bang but are not supposed to perform any significant role in natural surroundings nowadays. On the other hand, such heavyweight mesons are frequently produced in particle accelerator investigations in order to recognize the nature of the heavier kinds of quarks that comprise the heavier mesons. Since quarks have a spin 1/2, the transformation in quark number amongst mesons and baryons consequences in conservative two-quark mesons being bosons, while baryons are fermions. Each type of meson has an analogous antiparticle (antimeson) in which quarks are substituted by their analogous antiquarks and vice versa. For instance, a positive pion (π+) is made of one up quark and one down antiquark; and its corresponding antiparticle, the negative pion (π−), is made of one up antiquark and one down quark. Some significant properties of mesons are as follow [47].

- Composed of a quark and anti-quark
- All are very unstable
- They are not part of everyday matter
- Have a mass between that of the electron and the proton
- All decay into electrons, positrons, neutrinos and photons.

To recap, quarks form two main types of particles: (1) BARYONS, of which the proton and neutron are the least massive examples, are composed of three "valence quarks," one of each color. All baryons are fermions. (2) MESONS, which are a collection of a valence quark-antiquark pair. All mesons are bosons. All baryons and mesons (collectively called HADRONS) are colorless and held together by the strong interaction. Of all the hadrons, only the proton appears to be stable. Fig. (**9.16**) indicates an illustrative model for baryons and mesons.

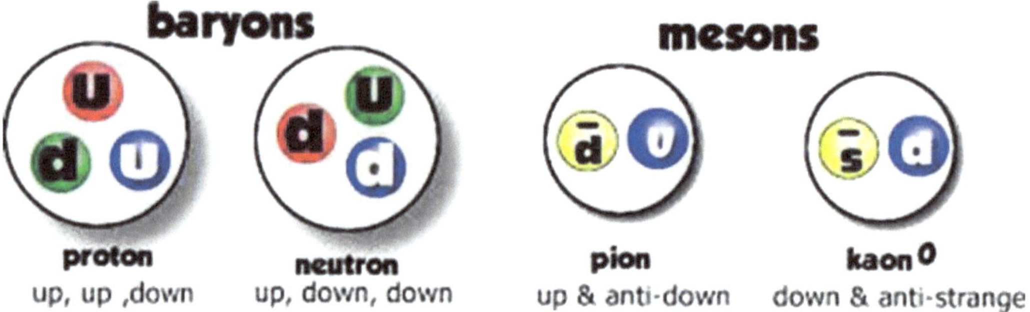

Fig. (9.16). Representative model for baryons and mesons. Image Credit: web2.ph.utexas.edu, stdmodel3021. Image Source: https://web2.ph.utexas.edu/~coker2/index.files/stdmodel.htm.

2.4.9. Fermions

A fermion is a particle that monitors Fermi–Dirac statistics and usually has half-odd integer spin: spin 1/2, spin 3/2, *etc*. Thesc particles follow the Pauli Exclusion Principle. Fermions consist of all quarks and leptons, as well as all combinations of particles made of an odd number of these, such as all baryons and numerous atoms and nuclei. Fermions are different from bosons, which follow Bose-Einstein statistics. Roughly fermions are elementary particles, such as the electrons, and some are complex particles, such as the protons. Agreeing to the spin-statistics theorem in relativistic quantum field theory, particles with numeral spin are bosons, whereas particles with half-integer spin are fermions. In tallying to the typical spin, fermions have additional specific property: they retain preserved baryon or lepton quantum numbers. Consequently, what is ordinarily denoted as the spin-statistics relation is, in information, a spin statistics-quantum number relation [48].

As a result of the Pauli Exclusion Principle, only one fermion can reside in a particular quantum state at a given time. If numerous fermions have the same three-dimensional probability distribution, then at least one property of each fermion, such as its spin, must be altered. Fermions are typically related to matter, while bosons are usually forced transporter particles, even though in the present state of particle physics, the difference between the two perceptions is uncertain. Inadequately relating fermions can also display bosonic performance under extreme circumstances. At low temperatures, fermions show superfluidity for uncharged particles and superconductivity for charged particles. Complex fermions, such as protons and neutrons, are the key building blocks of everyday matter. The name fermion was invented by English academic physicist Paul Dirac from the last name of Italian physicist Enrico Fermi [49]. Fig. (**9.17**): Elementary fermions: the particles of matter and quantum statistics between Bosons and Fermions.

In short,

- Fermions are particles that obey the Pauli Exclusion Principle
- A fermion is any particle that has a half-integer spin
- Ex. 1/2, 3/2, 5/2
- Quarks and leptons, as well as most composite particles, like protons and neutrons, are fermions.

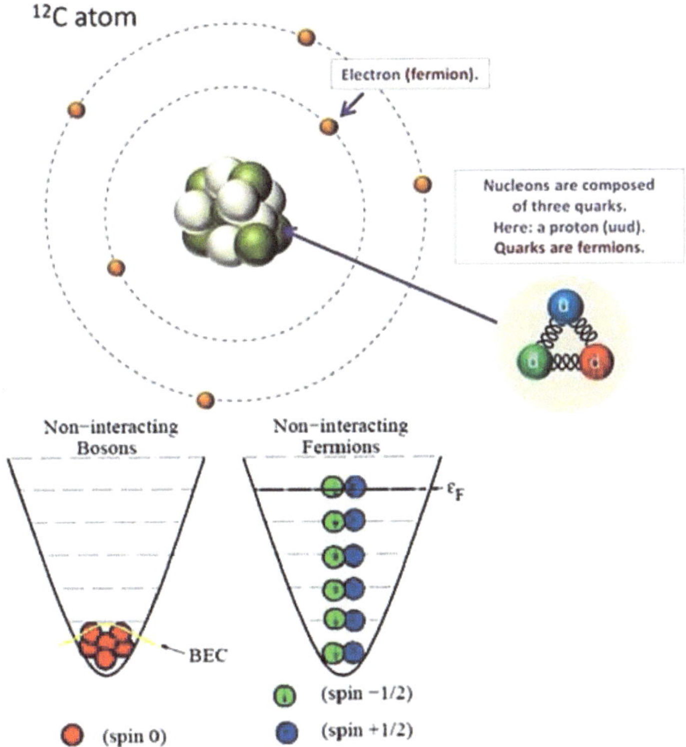

Fig. (9.17). Elementary fermions: the particles of matter. Image Credit and Source: Steemit, https://steemit.com/science/@muphy/elementary-fermions-the-particles-of-matter-particle-physics-series-episode-3a

2.4.10. Bosons

In quantum mechanics, a boson [50, 51] is a particle that trails Bose–Einstein statistics. Bosons make up one of two categories of elementary particles, the other being fermions [52]. The name boson was invented by Paul Dirac [53, 54] to remember the contribution of Satyendra Nath Bose, an Indian physicist and professor of physics at the University of Calcutta, in developing, Bose–Einstein statistics with Albert Einstein, which conceived the individualities of elementary particles [55].

Illustrations of bosons are fundamental particles such as photons, gluons, W and Z bosons, the freshly revealed Higgs boson, and the theoretical graviton of quantum gravity. Some complex particles are also bosons, such as mesons and stable nuclei of even mass numbers, such as deuterium. A significant representation of bosons is that there is no constraint on the number of them that live in the same quantum state. This property is demonstrated by helium-4 when it is cooled to grow a superfluid [56]. Unlike bosons, two indistinguishable fermions cannot reside in the same quantum state. However, the elementary particles that makeup matter (*i.e.*, leptons and quarks) are fermions; the elementary bosons are force carriers that function as the 'glue' holding systematized the matter [57]. These things hold for entire particles with integer spin (s = 0, 1, 2, *etc.*) as a result of the spin-statistics theorem. When a gas of Bose particles is cooled down to temperatures very near to absolute zero, then the kinetic energy of the particles reduces to an insignificant extent, and they reduce into the lowest energy level state. This state is called a Bose–Einstein condensate. This property is also the explanation for superfluidity.

Bosons may be either elementary, like photons, or composite, like mesons.

While most bosons are complex particles, in the Standard Model of Particle Physics, there are five elementary bosons:

The Standard Model requires (at least) one scalar boson (spin=0)

H^0 (Higgs boson)

The four-vector bosons (spin=1) that are the gauge bosons for the Standard Model:

γ Photon

g Gluons (eight different types)

Z Neutral weak boson

$W\pm$ Charged weak bosons (two types)

In Short:

- Bosons are particles that do not obey the Pauli Exclusion Principle
- All the force carrier particles are bosons, as well as those composite particles with an even number of fermion particles (like mesons).

- They have integer spins
- Ex. 0, 1, 2

Fig. (**9.18**) indicates the schematics of bosons and fermions.

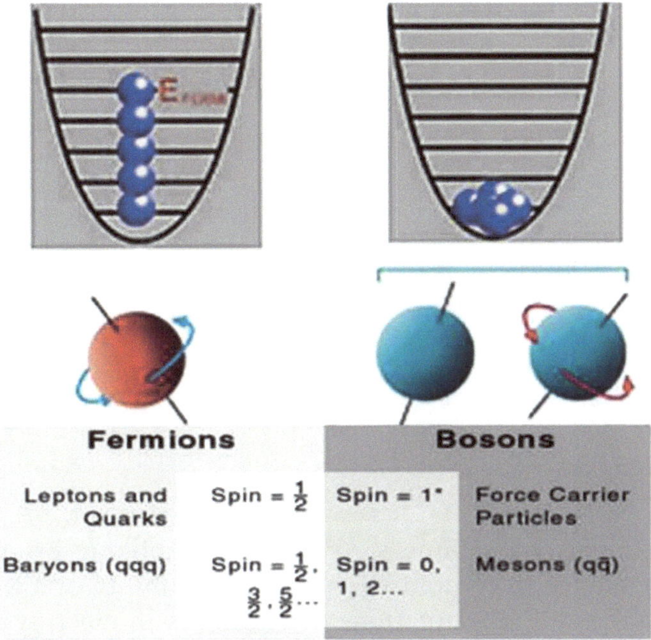

Fig. (9.18). The schematics of bosons and fermions.

3. GENERATIONS OF MATTER

In particle physics, a generation or family is a separation of elementary particles. Between generations, particles change by their flavour quantum number and mass, but their electric and strong interactions are equal. There are three generations rendering the Standard Model of particle physics (see Fig. (**9.7**)). Each generation encompasses two types of leptons and two types of quarks. The two leptons may be categorized into one with electric charge −1 (electron-like) and neutral (neutrino); the two quarks may be categorized into one with charge −1⁄3 (down-type) and one with charge +2⁄3 (up-type). The basic features of the quark-lepton generation, such as their masses and mixings *etc.*, can be defined by some of the anticipated family regularities. Fig. (**9.19**) reveals the three generations of matter.

- Mass increases from 1 generation to the next
- Going down in each generation, the charges are: +2/3, -1/3, 0, -1
- These are all in multiples of the elementary charge

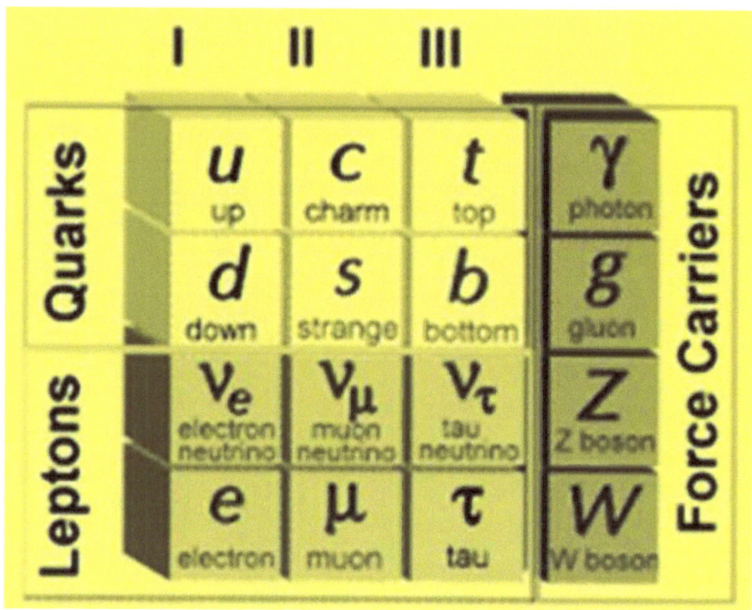

Fig. (9.19). Three Generations of Matter.

Fourth and additional generations are deliberated unlikely by many theoretical physicists. Some advice in contradiction to the probability of a fourth-generation are grounded on the subtle adaptations of exactness in electroweak observables that extra generations would induce; such modifications are strongly broken by measurements. Moreover, a fourth-generation with a "light" neutrino (one with a mass less than about 45 GeV/c2) has been ruled out by capacities of the decay widths of the Z boson at CERN's Large Electron–Positron Collider (LEP) [58]. However, explorations at high-energy colliders for particles from a fourth-generation carry on, but as of yet, no confirmation has been witnessed [59]. In such explorations, fourth-generation particles are signified by the same symbols as third-generation ones with an added prime (*e.g.*, b′ and t′). The lower bound for the fourth generation of quark (b′, t′) masses is currently at 1.4TeV from experiments at the LHC. Table **9.3** streamlines the three generations of matter.

4. FUNDAMENTAL FORCES

In physical science, the significant connections, also recognized as fundamental forces, are the exchanges that do not give the idea to be reducible to more elementary associations. There are four fundamental interactions identified to take place [60]: the gravitational and electromagnetic interactions, which collect noteworthy long-range forces whose possessions can be seen agreeably in routine life, and the strong and weak interactions, which yield forces at the microscopic level, subatomic regions and direct nuclear interactions. Some specialists postulate that a fifth force might occur, but these ideas are just enthusiasm and imaginary [61 - 63].

Table 9.3. Three generations of matter.

Type	Generations of matter		
	First	Second	Third
Quarks			
up-type	up	charm	top
down-type	down	strange	bottom
Leptons			
charged	electron	muon	tau
neutral	electron neutrino	muon neutrino	tau neutrino

Individually the recognized fundamental interactions can be characterized arithmetically as a field. The gravitational force is recommended for the twisting of space-time, well-defined by Einstein's general theory of relativity. The other three are discrete quantum fields, and their interactions are occurred by elementary particles marked by the standard model of particle physics [64]. Inside the Standard Model, the strong interaction is supported by a particle named the gluon and is responsible for quarks holding them connected to form hadrons, such as protons and neutrons. As a remaining outcome, it generates the nuclear force that impasses the concluding particles to form atomic nuclei. The weak interaction is reinforced by particles, so-called W and Z bosons, and also reveals the movements of the nucleus of atoms, settling down the radioactive decay. The electromagnetic force, approved by the photon, generates electric and magnetic fields, which are held responsible for the pull amongst orbital electrons and atomic nuclei, which clutches atoms together, as well as electromagnetic waves and chemical bonding with visible light, and forms the foundation for electrical technology. Even though the electromagnetic force is extremely stronger than gravity, it has a tendency to terminate itself out inside outsized entities, so over an astronomical or extremely large space gravity inclines to be the dominant force and is responsible for holding together the large-scale configurations in the universe, such as planets, stars, and galaxies.

Many academic physicists have faith in these fundamental forces to be connected and to become integrated into a distinct force at very high energies on a microscopic measure, the Planck measure, but particle accelerators cannot produce the massive energies essential to experimentally investigate this [65]. Formulating a mutual hypothetical structure that would clarify the relationship between the forces in a single model is possibly the highest goal of today's hypothetical physicists. The weak and electromagnetic forces have previously been incorporated with the electroweak theory of Sheldon Glashow, Abdus Salam, and Steven Weinberg, for which they were acknowledged for the 1979 Nobel Prize in physics [66 - 68]. Some physicists strived to bond the electroweak and strong fields inside what is called a Grand Unified Theory (GUT). An even greater experiment is to discover a technique to quantize the gravitational field, causing a model of quantum gravity (QG) that would unite gravity in a mutual theoretical framework with the other three forces. Some theories, particularly string theory, try to find both QG and GUT inside one background, combining all four fundamental interactions and mass generation within a theory of everything (ToE). These forces include interactions that are attractive or repulsive, decay and annihilation. There are four basic forces.

a. Strong
b. Weak
c. Electromagnetic
d. Gravitational

Fig. (**9.20**) reveals the schematic illustration of fundamental forces.

4.1. The Strong Forces

It is the strongest fundamental force, many times stronger than gravity (1038 times stronger: that's 1 trailed by 38 zeros). But it works only over very short distances of a few femtometres (fm). A femtometre is 10−15 (0.00000 00000 00001) meters. Scientists regularly think about the two ways the strong interaction works as distinct forces: the *colour force* and the *nuclear force*. At distances of 0.8 fm and less, the colour force grips subatomic particles like protons and neutrons composed. At distances of 1 to 3 fm, the remaining strong force is what holds onto protons and neutrons composed in the atomic nucleus, so it is termed the nuclear force. (This is like thinking of electricity and magnetism as detached forces when the fundamental force is electromagnetism).

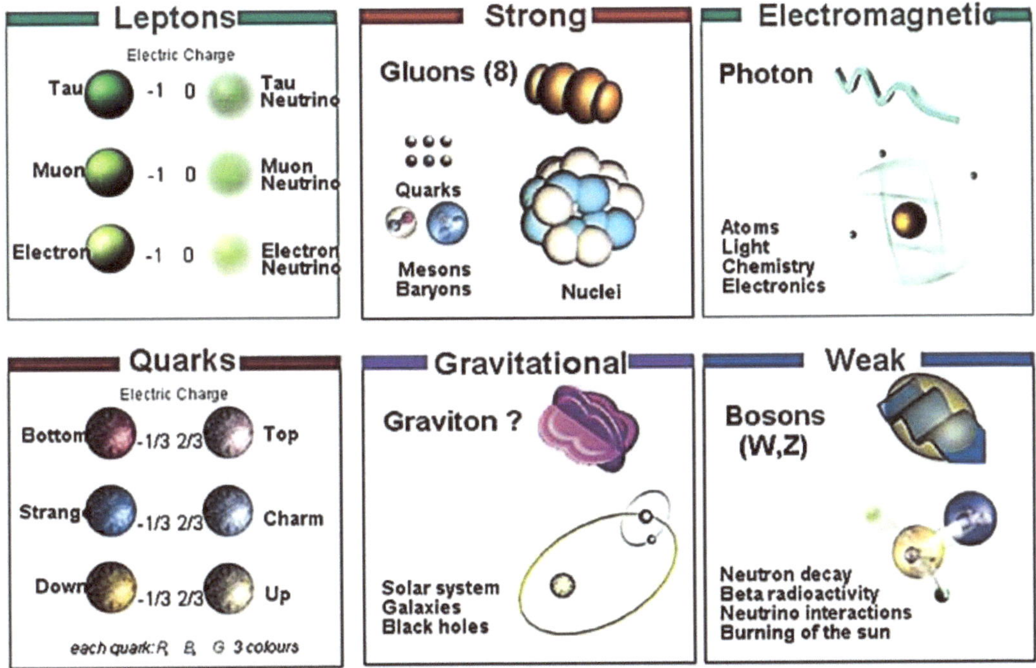

Fig. (9.20). The schematic illustration of fundamental forces. Image Source: http://mindblowing-physics.pbworks.com/f/1332432623/force%20particles.jpeg.

A force that can hold a nucleus together against the huge forces of repulsion of the protons is certainly strong. On the other hand, it is not an inverse square force like the electromagnetic force, and it has a very short range. Yukawa demonstrated the strong force as an exchange force in which the exchange particles are pions and other significant particles. The range of a particle exchange force is inadequate by the uncertainty principle. It is the strongest of the four fundamental forces.

As the protons and neutrons which make up the nucleus are themselves deliberated to be made up of quarks, and the quarks are well-thought-out to be held organized by the color force, the strong force between nucleons may be deliberated to be leftover color force. In the standard model, consequently, the elementary exchange particle is the gluon which enables the forces amongst quarks. Since the discrete gluons and quarks are enclosed in the interior of the proton or neutron, the masses recognized to them cannot be recycled in the range connection to expect the range of the force. When something is observed as coming out from a proton or neutron, then it must be essentially at least a quark-antiquark pair, so it is then reasonable that the pion as the lightest meson should function as an interpreter of the maximum range of the strong force amongst nucleons [69]. In short,

It is the strongest of the 4 forces.

Is only effective at distances less than 10^{-15} meters (about the size of the nucleus).

Holds quarks together.

This force is carried by gluons.

4.2. The Weak Forces

Weak interaction, the weak force or weak nuclear force [1], is one of the four fundamental forces in the cosmos. It is carried by particles known as the W and Z bosons, which are gauge bosons. The weak force origins beta decay, a form of radioactivity [2]. At enormously high energy levels, the force of weak interaction and electromagnetism initiate to act the same, called electroweak interaction.

Unique force amongst the four fundamental forces is weak interaction that involves the interchange of the intermediary vector bosons, the W and the Z. Subsequently, the mass of these particles is on the order of 80 GeV, and the uncertainty principle commands a range of about 10-18 meters which is about 0.1% of the diameter of a proton. The weak interaction gives rise to modifying one flavor quark into an additional quark. It is risky to the configuration of the universe as,

i. The sun would not burn without it since the weak interaction geneses the transmutation p -> n so that deuterium can form and deuterium fusion can proceed.
ii. It is indispensable for the accumulation of heavy nuclei.

The major role of the weak force in the alteration of quarks generates the interaction elaborate in numerous degenerations of nuclear particles, which prerequisite an alteration of a quark from one flavor to an additional quark. It was in radioactive decay, such as beta decay, that the occurrence of the weak interaction was primarily revealed. The weak interaction is the solitary phenomenon that monitors the advancement in which one quark can convert to another quark or a lepton to another lepton - the pretended as “flavor changes.” By discharging a W boson, one kind of fermion can show the modification into another kind: a down quark can spit out a W- and change into an up quark. Not any of the other forces can adapt to the individualism of the particles they work together with. The revolution of the W and Z particles in 1983 was addressed as an authentication of the replicas, which link the weak force to the electromagnetic force in an electroweak association. The weak interaction acts in the midst of both quarks and leptons, while the strong force does not act between leptons. “Leptons have no color, so they do not subsidize in the strong interactions; neutrinos have no charge, so they experience no electromagnetic forces, but all of them connect in the weak interactions [70].

4.3. Electromagnetic Forces

Amongst the four fundamental forces, the electromagnetic force exposes itself through the forces between charges (Coulomb's Law) and the magnetic force, both of which are summed up in the Lorentz force law. Fundamentally, both magnetic and electric forces are pointers of an exchange force regarding the exchange of photons. The quantum methodology to the electromagnetic force is authorized quantum electrodynamics or QED. The electromagnetic force is a force of infinite range which monitors the inverse square law and is of the indistinguishable form as the gravity force.

The electromagnetic force embraces atoms and molecules structured. In detail, the forces of electrical attraction and repulsion of electric charges are so prevailing over the other three fundamental forces that they can be well-organized to be irrelevant as determiners of atomic and molecular structure. Even magnetic extraordinary effects are characteristically apparent only at high determination and as slight expansions.

4.4. Gravitational Forces

Gravitation is a normal occurrence by which all belongings with mass or energy, including planets, stars, galaxies, and even light [71], are attracted to one another. Gravity has an infinite range. However, its special effects become weaker as entities get further away. Gravity is most accurately well-defined by the general theory of relativity (anticipated by Albert Einstein in 1915), which describes gravity not as a force but as a degree of masses moving together with geodesic appearances in a frizzy space-time created by the irregular scattering of mass. The black hole is one of the most thrilling illustrations of this curvature of space-time, from which nothing, not even light, can outflow once past the black hole's existence forecasts [72].

Gravity is the weakest of the four fundamental interactions of physics, about 1038 times weaker than the strong interaction, 1036 times weaker than the electromagnetic force and 1029 times weaker than the weak interaction. As a result, it has no important effect at the level of subatomic particles [73]. In contrast, it is the foremost interaction at the macroscopic measure and is the foundation of the development, shape and orbit or trajectory of planetary bodies.

Current demonstrations of particle physics put forward the fact of significant existence of gravity in the Universe, conceivably in the form of quantum gravity, supergravity or gravitational independence, along with regular space and time, established in 10−43 seconds after the birth of the Universe so-called as the Planck era, credibly commencing an original state, such as a virtual particle, quantum vacuum or false vacuum in a presently anonymous way. Efforts to progress a concept of gravity consistent with quantum mechanics, a quantum gravity theory, which would authorize gravity to be incorporated in a joint mathematical framework (a theory of everything) with the other three necessary physics interactions, are a present expanse of exploration.

Table **9.4** summarizes the fundamental forces, their strength, type and range of corresponding particles.

Table 9.4. The fundamental forces, their strength, type and range of corresponding particles.

Fundamental Force / Interaction	Strength	Range (m)	Particle
Strong (+) π (+) / (N) π (+) Force which holds nucleus together	1	10^{-15} (diameter of a medium sized nucleus)	gluons, π(nucleons)
Electro-magnetic ←(+) (+)→ / (−)→ ←(+)	$\frac{1}{137}$	Infinite	photon mass = 0 spin = 1
Weak ν→ W- neutrino interaction induces beta decay e⁻	10^{-6}	10^{-18} (0.1% of the diameter of a proton)	Intermediate vector bosons W^+, W^-, Z_0, mass > 80 GeV spin =1
Gravity (m)→ ←(m)	6×10^{-39}	Infinite	graviton ? mass = 0 spin = 2

REFERENCES

[1] Jeremy Butterfield, John Earman, Ed., Issues in the Philosophy of Cosmology *Philosophy of Physics (Handbook of the Philosophy of Science)* vol. 3. North Holland.

[2] "An Open Letter to the Scientific Community as published in New Scientist, May 22, 2004", Cosmologystatement.org. 1 April 2014. Archived from the original on 1 April 2014, 2014. An Open Letter to the Scientific Community as published in New Scientist, May 22, 2004". Retrieved 27 September.

[3] J. Beringer, J-F. Arguin, R.M. Barnett, K. Copic, O. Dahl, D.E. Groom, C-J. Lin, J. Lys, H.

Murayama, C.G. Wohl, W-M. Yao, P.A. Zyla, C. Amsler, M. Antonelli, D.M. Asner, H. Baer, H.R. Band, T. Basaglia, C.W. Bauer, J.J. Beatty, V.I. Belousov, E. Bergren, G. Bernardi, W. Bertl, S. Bethke, H. Bichsel, O. Biebel, E. Blucher, S. Blusk, G. Brooijmans, O. Buchmueller, R.N. Cahn, M. Carena, A. Ceccucci, D. Chakraborty, M-C. Chen, R.S. Chivukula, G. Cowan, G. D'Ambrosio, T. Damour, D. de Florian, A. de Gouvêa, T. DeGrand, P. de Jong, G. Dissertori, B. Dobrescu, M. Doser, M. Drees, D.A. Edwards, S. Eidelman, J. Erler, V.V. Ezhela, W. Fetscher, B.D. Fields, B. Foster, T.K. Gaisser, L. Garren, H-J. Gerber, G. Gerbier, T. Gherghetta, S. Golwala, M. Goodman, C. Grab, A.V. Gritsan, J-F. Grivaz, M. Grünewald, A. Gurtu, T. Gutsche, H.E. Haber, K. Hagiwara, C. Hagmann, C. Hanhart, S. Hashimoto, K.G. Hayes, M. Heffner, B. Heltsley, J.J. Hernández-Rey, K. Hikasa, A. Höcker, J. Holder, A. Holtkamp, J. Huston, J.D. Jackson, K.F. Johnson, T. Junk, D. Karlen, D. Kirkby, S.R. Klein, E. Klempt, R.V. Kowalewski, F. Krauss, M. Kreps, B. Krusche, Y.V. Kuyanov, Y. Kwon, O. Lahav, J. Laiho, P. Langacker, A. Liddle, Z. Ligeti, T.M. Liss, L. Littenberg, K.S. Lugovsky, S.B. Lugovsky, T. Mannel, A.V. Manohar, W.J. Marciano, A.D. Martin, A. Masoni, J. Matthews, D. Milstead, R. Miquel, K. Mönig, F. Moortgat, K. Nakamura, M. Narain, P. Nason, S. Navas, M. Neubert, P. Nevski, Y. Nir, K.A. Olive, L. Pape, J. Parsons, C. Patrignani, J.A. Peacock, S.T. Petcov, A. Piepke, A. Pomarol, G. Punzi, A. Quadt, S. Raby, G. Raffelt, B.N. Ratcliff, P. Richardson, S. Roesler, S. Rolli, A. Romaniouk, L.J. Rosenberg, J.L. Rosner, C.T. Sachrajda, Y. Sakai, G.P. Salam, S. Sarkar, F. Sauli, O. Schneider, K. Scholberg, D. Scott, W.G. Seligman, M.H. Shaevitz, S.R. Sharpe, M. Silari, T. Sjöstrand, P. Skands, J.G. Smith, G.F. Smoot, S. Spanier, H. Spieler, A. Stahl, T. Stanev, S.L. Stone, T. Sumiyoshi, M.J. Syphers, F. Takahashi, M. Tanabashi, J. Terning, M. Titov, N.P. Tkachenko, N.A. Törnqvist, D. Tovey, G. Valencia, K. van Bibber, G. Venanzoni, M.G. Vincter, P. Vogel, A. Vogt, W. Walkowiak, C.W. Walter, D.R. Ward, T. Watari, G. Weiglein, E.J. Weinberg, L.R. Wiencke, L. Wolfenstein, J. Womersley, C.L. Woody, R.L. Workman, A. Yamamoto, G.P. Zeller, O.V. Zenin, J. Zhang, R-Y. Zhu, G. Harper, V.S. Lugovsky, and P. Schaffner, "Review of Particle Physics", *Phys. Rev. D Part. Fields Gravit. Cosmol.,* vol. 86, no. 1, p. 010001, 2012. [http://dx.doi.org/10.1103/PhysRevD.86.010001]

[4] The Nobel Prize in Physics 2011. NobelPrize.org. Nobel Prize Outreach AB 2023. Thu. 12 Jan 2023. https://www.nobelprize.org/prizes/physics/2011/summary/

[5] S. Braibant, G. Giacomelli, and M. Spurio, *Particles and Fundamental Interactions: An introduction to particle physics.* 2nd ed. Springer, 2012. [http://dx.doi.org/10.1007/978-94-007-2464-8]

[6] F. Englert, The BEH Mechanism and its Scalar Boson. *Annalen der Physik.* 30. 2014. [http://dx.doi.org/10.1002/andp.201400881]

[7] S. Braibant, G. Giacomelli, and M. Spurio, *Particles and Fundamental Interactions: An Introduction to Particle Physics.* Springer, 2009, pp. 313-314.

[8] V. Cilek, Ed., *"Cosmic Influences on the Earth". Earth System: History and Natural Variability. I.* Eolss Publishers, 2009, p. 165.

[9] M. Ackermann, M. Ajello, A. Allafort, L. Baldini, J. Ballet, G. Barbiellini, M.G. Baring, D. Bastieri, K. Bechtol, R. Bellazzini, R.D. Blandford, E.D. Bloom, E. Bonamente, A.W. Borgland, E. Bottacini, T.J. Brandt, J. Bregeon, M. Brigida, P. Bruel, R. Buehler, G. Busetto, S. Buson, G.A. Caliandro, R.A. Cameron, P.A. Caraveo, J.M. Casandjian, C. Cecchi, O. Celik, E. Charles, S. Chaty, R.C.G. Chaves, A. Chekhtman, C.C. Cheung, J. Chiang, G. Chiaro, A.N. Cillis, S. Ciprini, R. Claus, J. Cohen-Tanugi, L.R. Cominsky, J. Conrad, S. Corbel, S. Cutini, F. D'Ammando, A. de Angelis, F. de Palma, C.D. Dermer, E. do Couto e Silva, P.S. Drell, A. Drlica-Wagner, L. Falletti, C. Favuzzi, E.C. Ferrara, A. Franckowiak, Y. Fukazawa, S. Funk, P. Fusco, F. Gargano, S. Germani, N. Giglietto, P. Giommi, F. Giordano, M. Giroletti, T. Glanzman, G. Godfrey, I.A. Grenier, M.H. Grondin, J.E. Grove, S. Guiriec, D. Hadasch, Y. Hanabata, A.K. Harding, M. Hayashida, K. Hayashi, E. Hays, J.W. Hewitt, A.B. Hill, R.E. Hughes, M.S. Jackson, T. Jogler, G. Jóhannesson, A.S. Johnson, T. Kamae, J. Kataoka, J. Katsuta, J. Knödlseder, M. Kuss, J. Lande, S. Larsson, L. Latronico, M. Lemoine-Goumard, F. Longo, F. Loparco, M.N. Lovellette, P. Lubrano, G.M. Madejski, F. Massaro, M. Mayer, M.N. Mazziotta, J.E. McEnery, J. Mehault, P.F. Michelson, R.P. Mignani, W. Mitthumsiri, T. Mizuno, A.A. Moiseev, M.E. Monzani, A. Morselli, I.V. Moskalenko, S. Murgia, T. Nakamori, R. Nemmen, E. Nuss, M. Ohno, T.

Ohsugi, N. Omodei, M. Orienti, E. Orlando, J.F. Ormes, D. Paneque, J.S. Perkins, M. Pesce-Rollins, F. Piron, G. Pivato, S. Rainò, R. Rando, M. Razzano, S. Razzaque, A. Reimer, O. Reimer, S. Ritz, C. Romoli, M. Sánchez-Conde, A. Schulz, C. Sgrò, P.E. Simeon, E.J. Siskind, D.A. Smith, G. Spandre, P. Spinelli, F.W. Stecker, A.W. Strong, D.J. Suson, H. Tajima, H. Takahashi, T. Takahashi, T. Tanaka, J.G. Thayer, J.B. Thayer, D.J. Thompson, S.E. Thorsett, L. Tibaldo, O. Tibolla, M. Tinivella, E. Troja, Y. Uchiyama, T.L. Usher, J. Vandenbroucke, V. Vasileiou, G. Vianello, V. Vitale, A.P. Waite, M. Werner, B.L. Winer, K.S. Wood, M. Wood, R. Yamazaki, Z. Yang, and S Zimmer, "Detection of the characteristic pion-decay signature in supernova remnants", *Science,* vol. 339, no. 6121, pp. 807-811, 2013.
[http://dx.doi.org/10.1126/science.1231160] [PMID: 23413352]

[10] Sarmistha Basu, "A Review work on Atmospheric Nanoparticles and Cosmic Ray activity are the cause of Climate Change and Biological Disorder of Public Health", *IOSR J. Appl. Phys*, vol. 12, no.4, pp. 44-49,2020.
[http://dx.doi.org/10.9790/4861-1204024449]

[11] Abramowski AAharonian FBenkhali F *et al.*, "Acceleration of petaelectronvolt protons in the Galactic Centre", *Nature* vol. 531, no. 7595, pp. 476-478, 2016.
[http://dx.doi.org/10.1038/nature17147]

[12] R.A. Mewaldt, *Cosmic Rays.* California Institute of Technology, 1996.

[13] L. Koch, J.J. Engelmann, P. Goret, E. Juliusson, N. Petrou, Y. Rio, A. Soutoul, B. Byrnak, N. Lund, and B. Peters, "The relative abundances of the elements scandium to manganese in relativistic cosmic rays and the possible radioactive decay of manganese 54", *Astron. Astrophys.,* vol. 102, no. 11, p. L9, 1981.

[14] Sloan T. and Wolfendale A. W., Cosmic rays, solar activity and the climate. *Environ. Res. Lett.* 8 045022 (IOP Publishing) 2013. [http://dx.doi.org/10.1088/1748-9326/8/4/045022]

[15] I. Morison, Introduction to Astronomy and Cosmology.*Extreme Space Weather Events* John Wiley & Sons: National Geophysical Data Center, 2008, p. 198.

[16] Benítez N., Maíz-Apellániz J., and Canelles M., "Evidence for Nearby Supernova Explosions", *Phys. Rev. Lett.* vol. 88, 2002.
[http://dx.doi.org/10.1103/PhysRevLett.88.081101]

[17] L. Fimiani, D.L. Cook, T. Faestermann, J.M. Gómez-Guzmán, K. Hain, G. Herzog, K. Knie, G. Korschinek, P. Ludwig, J. Park, R.C. Reedy, and G. Rugel, "Interstellar Fe 60 on the Surface of the Moon", *Phys. Rev. Lett.,* vol. 116, no. 15, p. 151104, 2016.
[http://dx.doi.org/10.1103/PhysRevLett.116.151104] [PMID: 27127953]

[18] S. Braibant, G. Giacomelli, and M. Spurio, *Particles and Fundamental Interactions: An Introduction to Particle Physics.* 1st ed. Springer, 2012, p. 1.
[http://dx.doi.org/10.1007/978-94-007-2464-8]

[19] V. Khachatryan, "(CMS Collaboration) “Observation of a new boson at a mass of 125 GeV with the CMS experiment at the LHC", *Phys. Lett. B,* p. 716, 2012.

[20] T. Abajyan, "(ATLAS Collaboration) (2012). “Observation of a new particle in the search for the Standard Model Higgs boson with the ATLAS detector at the LHC", *Phys. Lett. B,* vol. 716, pp. 1-29, 2012.
[http://dx.doi.org/10.1016/j.physletb.2012.08.020]

[21] Rühr F., Prospects for BSM searches at the high-luminosity LHC with the ATLAS detector, Nuclear and Particle Physics Proceedings,Volumes 273–275, 2016, Pages 625-630, ISSN 2405-6014, (https://www.sciencedirect.com/science/article/pii/S2405601415005830)
[http://dx.doi.org/10.1016/j.nuclphysbps.2015.09.094]

[22] *Unsolved mysteries: Supersymmetry* Berkeley Lab, 2013.*The Particle Adventure.* Berkeley Lab, 2013.

[23] *Charting the Course for Elementary Particle Physics.* National Academies Press, 2006, p. 68.

[24] R. W. Robinett, "Searching for supersymmetry in high-energy cosmic-ray interactions", *Physical Review D*, 33(5), 1239-1246. 1986. [http://dx.doi.org/10.1103/PhysRevD.33.1239]

[25] L. Smolin, "Atoms of Space and Time", *Sci. Am.,* vol. 16, no. 1, pp. 82-92, 2006. [http://dx.doi.org/10.1038/scientificamerican0206-82sp] [PMID: 14682040]

[26] Elisa Minucci, Searches for Lepton Flavour and Lepton Number violation in K+ decays, Nuclear and Particle Physics Proceedings, Volumes 318–323, 2022, Pages 165-169, ISSN 2405-6014,(https://www.sciencedirect.com/science/article/pii/S240560142200102X) [http://dx.doi.org/10.1016/j.nuclphysbps.2022.11.002]

[27] R. Nave, Leptons *HyperPhysics.* Georgia State: Georgia State University, Department of Physics and Astronomy., 2010.

[28] R. Nave, Quarks. *HyperPhysics.* Georgia State University, Department of Physics and Astronomy., 2008.

[29] S.S.M. Wong, *Introductory Nuclear Physics.* 2nd ed. Wiley Interscience, 1998, p. 30. [http://dx.doi.org/10.1002/9783527617906]

[30] K.A. Peacock, *The Quantum Revolution.* Greenwood Publishing Group, 2008, p. 125.

[31] M. Munowitz, *Knowing.* Oxford University Press, 2005, p. 35.

[32] Choi, and Q. Charles, "Does Antimatter Fall Up or Down? New Device May Provide Answer", *HuffingtonPost.com. N.p.,* 2014.

[33] R. Penrose, The mass of the classical vacuum.*The Philosophy of Vacuum.,* S. Saunders, H.R. Brown, Eds., Oxford University Press, 1991, pp. 21-26.

[34] M. Hertzog *et al.*, "Strong light–matter interactions: A new direction within chemistry", *Chem. Soc. Rev.,* 48(3):937–961, 2019. [http://dx.doi.org/10.1039/C8CS00193F]

[35] L. Amendola, S. Appleby, A. Avgoustidis, D. Bacon, T. Baker, M Baldi, N. Bartolo, A. Blanchard, C. Bonvin, S.Borgani, E. Branchini, C. Burrage, S. Camera, C.Carbone, L.Casarini, M. Cropper, C. de Rham, J. P. Dietrich, C. D. Porto, R. Durrer, A. Ealet, P. G. Ferreira, F. Finelli, J. García-Bellido, The Euclid Theory Working Group., "Cosmology and fundamental physics with the Euclid satellite", *Living Rev. Relativ.,* 21:2, 2018. [http://dx.doi.org/10.1007/s41114-017-0010-3]

[36] H. Agakishiev, MM, Aggarwal, Z. Ahammed, AV. Alakhverdyants, I. Alekseev, J. Alford *et al.* Observation of the antimatter helium-4 nucleus. *Nature.* 2011 May 19; 473(7347):353-356. [http://dx.doi.org/10.1038/nature10079]

[37] L. Canetti, M. Drewes, and M Shaposhnikov, "Matter and antimatter in the universe", *New J. Phys.,* vol. 14, no. 9, p. 095012, 2012. [http://dx.doi.org/10.1088/1367-2630/14/9/095012]

[38] E.J. Eichten, K.D. Lane, and M.E. Peskin, "New Tests for Quark and Lepton Substructure", *Phys. Rev. Lett.,* vol. 50, no. 11, pp. 811-814, 1983. [http://dx.doi.org/10.1103/PhysRevLett.50.811]

[39] "Fundamental Physical Constants from NIST", *The NIST Reference on Constants, Units, and Uncertainty.* US National Institute of Standards and Technology.

[40] M. Wurm, F. von, G. Franz *et al.*, (2010). Spectroscopy of Solar Neutrinos 203-214. [http://dx.doi.org/10.1002/9783527634842.ch13]

[41] M. Gell-Mann, "A schematic model of baryons and mesons", *Phys. Lett.,* vol. 8, no. 3, pp. 214-215, 1964. [http://dx.doi.org/10.1016/S0031-9163(64)92001-3]

[42] S.K. Choi *et al.* (2008). "Observation of a Resonancelike Structure in the π + − ψ ′ Mass Distribution in Exclusive B → K π + − ψ ′ Decays." *Physical review letters*. 100. 142001. [http://dx.doi.org/10.1103]

[43] R. Aaij *et al.*, "Observation of the Resonant Character of the Z(4430)- State", *Phys. Rev. Lett.*, 112(22) 222002. 2014. [http://dx.doi.org/10.11032]

[44] Baryon. *Encyclopedia Britannica* Encyclopedia Britannica, Inc., 2011.

[45] T. Nakano, and K. Nishijima, "Charge Independence for V -particles", *Prog. Theor. Phys.*, vol. 10, no. 5, pp. 581-582, 1953. [http://dx.doi.org/10.1143/PTP.10.581]

[46] C. Fowler, D. Griffiths, & W. de Groat, "The neural control of micturition". *Nat Rev Neurosci* 9, 453–466. 2008. [http://dx.doi.org/10.1038/nrn2401]

[47] Aubert, Jean-Jacques, U. Becker, P.J. *et al.* "Experimental observation of a heavy particle J." *Physical Review Letters* 33, no. 23 1974: 1404. [http://dx.doi.org/10.1103/PhysRevLett.33.14041]

[48] R.M Weiner, "Spin-statistics-quantum number connection and supersymmetry", *Phys. Rev. D Part. Fields Gravit. Cosmol.*, vol. 87, no. 5, pp. 055003-055005, 2013. [http://dx.doi.org/10.1103/PhysRevD.87.055003]

[49] *Notes on Dirac's lecture Developments in Atomic Theory at Le Palais de la Découverte,* 1945. *Uknatarchi dirac papers,* 1945.

[50] J.C. Wells, *Longman pronunciation dictionary.* Longman: Harlow, England, 1990. [http://dx.doi.org/10.1177/003368829002100208]

[51] E.P Gross, "Classical theory of boson wave fields", *Annals of Physics*, vol. 4, no.1, pp.57-74,1958. [http://dx.doi.org/10.1016/0003-4916(58)90037-X. ISSN 0003-4916]

[52] S. Carroll, Dark Matter, Dark Energy: The dark side of the universe. *Guidebook.* The Teaching Company., 2007, p. 43.

[53] Gnadig, Z. Kunszt, P. Hasenfratz, J. Kuti,Dirac's extended electron model, *Annals of Physics*, Vol 116, no. 2, 1978, Pages 380-407, ISSN 0003-4916. [http://dx.doi.org/10.1016/0003-4916(78)90238-5]

[54] Graham Farmelo, The Strangest Man: The Hidden Life of Paul Dirac, Mystic of the Atom. *Basic Books.*, 2009, p. 331.

[55] K. Daigle, *India: Enough about Higgs, let's discuss the boson,* Associated Press., 2012.

[56] Chatrchyan, S Khachatryan, VSirunyan, A M, "Observation of a new boson at a mass of 125 GeV with the CMS experiment at the LHC", *Physics Letters B, Vol 716*, Issue 1, 2012, Pages 30-61, ISSN 0370-2693. https://doi.org/10.1016/j.physletb.2012.08.021. (https://www.sciencedirect.com/science/article/pii/S0370269312008581)

[57] Sean Carroll, Explain it in 60 seconds: Bosons. *Symmetry Magazine.* Fermilab/SLAC.

[58] D. Decamp, "Determination of the number of light neutrino species", *Phys. Lett. B,* vol. 231, no. 4, pp. 519-529, 1989. [http://dx.doi.org/10.1016/0370-2693(89)90704-1]

[59] C. Amsler, M. Doser, M. Antonelli, D.M. Asner, K.S. Babu, H. Baer, H.R. Band, R.M. Barnett, E. Bergren, J. Beringer, G. Bernardi, W. Bertl, H. Bichsel, O. Biebel, P. Bloch, E. Blucher, S. Blusk, R.N. Cahn, M. Carena, C. Caso, A. Ceccucci, D. Chakraborty, M-C. Chen, R.S. Chivukula, G. Cowan, O. Dahl, G. D'Ambrosio, T. Damour, A. de Gouvêa, T. DeGrand, B. Dobrescu, M. Drees, D.A. Edwards, S. Eidelman, V.D. Elvira, J. Erler, V.V. Ezhela, J.L. Feng, W. Fetscher, B.D. Fields, B.

Foster, T.K. Gaisser, L. Garren, H-J. Gerber, G. Gerbier, T. Gherghetta, G.F. Giudice, M. Goodman, C. Grab, A.V. Gritsan, J-F. Grivaz, D.E. Groom, M. Grünewald, A. Gurtu, T. Gutsche, H.E. Haber, K. Hagiwara, C. Hagmann, K.G. Hayes, J.J. Hernández-Rey, K. Hikasa, I. Hinchliffe, A. Höcker, J. Huston, P. Igo-Kemenes, J.D. Jackson, K.F. Johnson, T. Junk, D. Karlen, B. Kayser, D. Kirkby, S.R. Klein, I.G. Knowles, C. Kolda, R.V. Kowalewski, P. Kreitz, B. Krusche, Y.V. Kuyanov, Y. Kwon, O. Lahav, P. Langacker, A. Liddle, Z. Ligeti, C-J. Lin, T.M. Liss, L. Littenberg, J.C. Liu, K.S. Lugovsky, S.B. Lugovsky, H. Mahlke, M.L. Mangano, T. Mannel, A.V. Manohar, W.J. Marciano, A.D. Martin, A. Masoni, D. Milstead, R. Miquel, K. Mönig, H. Murayama, K. Nakamura, M. Narain, P. Nason, S. Navas, P. Nevski, Y. Nir, K.A. Olive, L. Pape, C. Patrignani, J.A. Peacock, A. Piepke, G. Punzi, A. Quadt, S. Raby, G. Raffelt, B.N. Ratcliff, B. Renk, P. Richardson, S. Roesler, S. Rolli, A. Romaniouk, L.J. Rosenberg, J.L. Rosner, C.T. Sachrajda, Y. Sakai, S. Sarkar, F. Sauli, O. Schneider, D. Scott, W.G. Seligman, M.H. Shaevitz, T. Sjöstrand, J.G. Smith, G.F. Smoot, S. Spanier, H. Spieler, A. Stahl, T. Stanev, S.L. Stone, T. Sumiyoshi, M. Tanabashi, J. Terning, M. Titov, N.P. Tkachenko, N.A. Törnqvist, D. Tovey, G.H. Trilling, T.G. Trippe, G. Valencia, K. van Bibber, M.G. Vincter, P. Vogel, D.R. Ward, T. Watari, B.R. Webber, G. Weiglein, J.D. Wells, M. Whalley, A. Wheeler, C.G. Wohl, L. Wolfenstein, J. Womersley, C.L. Woody, R.L. Workman, A. Yamamoto, W-M. Yao, O.V. Zenin, J. Zhang, R-Y. Zhu, P.A. Zyla, G. Harper, V.S. Lugovsky, and P. Schaffner, "Review of Particle Physics", *Phys. Lett. B,* vol. 667, no. 1-5, pp. 1-6, 2008. [http://dx.doi.org/10.1016/j.physletb.2008.07.018]

[60] Raibant Sylvie, Giacomelli Giorgio, and Spurio Maurizio, Particles and Fundamental Interactions: An Introduction to Particle Physics (illustrated Ed.). *Springer Science & Business Media.*, 2011, p. 109.

[61] O. Fackler, and J.T. Van Tran, *5th Force Neutrino Physics.* Atlantica Séguier Frontières, 1988.

[62] E.W. Weisstein, Fifth Force. *World of Science.* Wolfram Research, 2007.

[63] A. Franklin, and E. Fischbach, *The Rise and Fall of the Fifth Force: Discovery, Pursuit, and Justification in Modern Physics.* 2nd ed. Springer, 2016. [http://dx.doi.org/10.1007/978-3-319-28412-5]

[64] *The Standard Model of Particle Physics | symmetry magazine.* Available from: www.symmetry magazine.org

[65] Rashmi Shivni, The Planck scale.*Symmetry magazine.* Fermilab/SLAC., 2016.

[66] "The Nobel Prize in Physics 1979", *NobelPrize.org.*

[67] "The Nobel Prize in Physics 1979", *NobelPrize.org.*

[68] "The Nobel Prize in Physics 1979", *NobelPrize.org.*

[69] D.J. Griffiths, *Introduction to Elementary Particles.* John Wiley & Sons, 1987. [http://dx.doi.org/10.1002/9783527618460]

[70] Griffiths, D. (2008) Introduction to Elementary Particles. 2nd Edition 2008, Wiley-VCH, Weinheim. ISBN: 978-3-527-40601-2.

[71] T. Duncan, *Advanced physics for Hong Kong: I. Mechanics & Electricity.* John Murray, 1995.

[72] J.B. Barbour, and H. Pfister, *Mach's Principle: From Newton's Bucket to Quantum Gravity.* Springer Science & Business Media, 1995, p. 69.

[73] R.M. Wald, *General Relativity.* University of Chicago Press, 1984. [http://dx.doi.org/10.7208/chicago/9780226870373.001.0001]

CHAPTER 10

Nuclear Astrophysics

Abstract: Nuclear Astrophysics is a field at the joining of nuclear physics and astrophysics that try to find and recognize how nuclear processes shape the universe. In essence, we look in the present chapter for the connection between the properties of atomic nuclei and the properties of planets, stars, and galaxies. The present chapter encloses astrophysics of the universe, theories on the creation of the universe, formation of a star, thermonuclear reactions in stars, nucleosynthesis, a process for the production of elements, death of a star and the age of our Galaxy.

Keywords: Galaxy, Nuclcar Astrophysics, Nucleosynthesis, Star, Universe.

1. INTRODUCTION

Nuclear astrophysics is an interdisciplinary part of both nuclear physics and astrophysics, containing a close relationship between investigators in numerous subfields of each field. This comprises, particularly, nuclear reactions and their proportions as they take place in cosmic surroundings, and demonstrating of cosmological entities where these nuclear reactions may take place, but also deliberations of cosmic advancement of isotopic and chemical evolution to identify the fundamental structures. Limitations from interpretations include multiple messengers, all across the electromagnetic spectrum (nuclear gamma-rays, X-rays, optical, and radio/sub-mm astronomy), as well as isotopic dimensions of solar-system constituents such as meteorites and their stardust presences, cosmic rays, material deposits on Earth and Moon). Nuclear physics experiments address stability (*i.e.*, lifetimes and masses) for atomic nuclei well outside the system of stable nuclides into the territory of unstable or radioactive nuclei, practically to the parameters of certain nuclei, and below up to neutron star matter and high temperatures plasma up to 10^9 K [1].

Models and their principles are vital in this domain, as cosmic nuclear reaction surroundings cannot be understood but, at finest, moderately moved toward by experiments. In overall expressions, nuclear astrophysics purposes of recognizing the foundation of the isotopes and chemical components, and the role of nuclear energy generation, in cosmic foundations such as stars, novae, supernovae, and intense binary-star connections.

Ritesh Kohale, Sanjay J. Dhoble & Vibha Chopra

Nuclear astrophysics leftovers are a difficult mystery to solve in science [2]. The existing agreement on the origins of isotopes and elements is that only helium and hydrogen (and traces of lithium, beryllium, and boron) can be molded in an identical Big Bang, while all other isotopes and elements there are twisted in cosmic matters that formed far ahead, such as in stars and their explosions. Even though the grounds of nuclear astrophysics give a conceivable and clear idea, many problems remain unanswered. One example from nuclear reaction physics is helium fusion (specifically the ^{12}C (α, γ) ^{16}O reaction(s)) [3]; others are the cosmological site of the r-process, irregular lithium abundances in Residents III stars, and the bang mechanism in core-collapse supernovae and the originators of thermonuclear supernovae. Contemporary explanations of the cosmic progression of fundamental abundances are approximately reliable to those detected in the galaxy and Solar System, whose distribution duration is one trillion in twelve orders of magnitude.

1.1. Astrophysics of the Universe

Astrophysics is a science that employs the approaches and main beliefs of physics in studying astrophysical entities and singularities [4, 5]. In the middle of the subjects studied are the galaxies, the Sun, extrasolar planets, the interstellar medium, other stars and the cosmic microwave background [6]. Emissions from these entities are scrutinized across all parts of the electromagnetic spectrum, and the properties examined contain chemical composition, temperature and luminosity. Because astrophysics is a very comprehensive subject, astrophysicists relate concepts and methods from numerous disciplines of physics, including electromagnetism, statistical mechanics, classical mechanics, quantum mechanics, relativity, thermodynamics, nuclear and particle physics, and atomic and molecular physics [7].

The Universe is vast, far bigger than we can imagine or comprehend. It is enormous and includes Earth and the remainder of space, the sun, the moon, other planets, stars and galaxies. The Greece Philosopher Aristotle, in 300s B.C., reasoned that a perfectly spherical earth had to be at the center of the Universe. Later, Ptolemy summed up that Earth was at the center, and all the other heavenly bodies revolve around it. Later Copernicus put forward his idea that the motion of heavenly bodies could be better explained if the sun and nor earth were at the centre of things. The later development of the telescope proved the concept that the sun was just one of many stars. Earth was not the centre of the Universe, nor the sun, nor even the Milky Way. It became apparent that our Milky Way galaxy was one of the many galaxies. Now the problem was to measure the distance of those off Galaxies. Edwin Hubble proposed that, since the founder galaxies had the larger redshifts, the velocity at which they receded was proportional to their distance. It had a deep meaning that the Universe was expanding. Further, it implied that if the galaxies were receding, they must have been concentrated into a small region of space at some time, meaning that the Universe must have a definite age. This can simply be expressed by Hubble's law. It relates the galaxy's distance to its recession velocity *via* a value known as Hubble's constant. The Hubble's constant was considered to be a direct measure of the universe's age if it is expanding at a constant rate. But the problem is that measuring Hubble's constants was not straightforward, which eventually led to the discovery of some other theories to explain the creation of the universe.

1.2. Theories on the Creation of the Universe

There are three well-known theories explaining the creation of the universe. These are the Big Bang Theory, the Oscillation Theory and the Steady-State Theory. None of these theories can completely explain the phenomena of creation, evolution and the death of the universe.

1.2.1. The Big Bang Theory

According to *George Gamow,* the universe came into existence nearly 15 billion years ago from a sudden explosion and the phenomenon is known as Big Bang. In this phenomenon, the entire matter, energy, space and time were created simultaneously. It is believed that all of a sudden, in one in a million fractions of section (10^{-12}s), the universe expanded to a size equal to the present size of our solar system. At that stage, the universe had a temperature of 10^{13} K. The basic concept of the Big Bang theory is that from the beginning of the creation of the universe, it expanded continuously and correspondingly cooled down [8].

The Big Bang theory explains the evolution of the Universe through different stages. It is theoretically assumed that from the beginning of time until 10^{-43}s, known as Planck time, all the four fundamental forces, namely Gravitational Force, Electromagnetic force, Nuclear Strong force and Weak force, jointly acted as a single unified force. At the end of the 10^{-43}s, the Gravitation force separated from the other forces. At that time, the temperature of the Universe was 10^{32} K. This Universe then suddenly expanded and thereby cooled down. At that time, there was not even a single particle of matter present in the Universe; it was all energy. When the Universe was 10^{-35} sec old, its temperature reduced to 10^{28} K due to cooling, and the Strong Nuclear force started separating from the remaining other forces. Further, when the age of the Universe was between 10^{-35}s and 10^{-32}s, the size of the universe got doubled and the temperature reduced to 10^{20} K; the strong nuclear forces and electromagnetic forces got separated from each other [9]. Further, there was fluxation of particles, especially quarks, antiquarks, photons and electrons, from the vast energy reserve of the Universe. When the age of the Universe was 10^{-12} s, it was as big as the present size of our solar system. When the age was 10^{-9}s, the temperature was lowered to 10^{14} K, then the Electromagnetic force and Weak forces separated. When the age of the Universe was 10^{-6}s and the temperature 10^{13}K, the quarks started combining to form hadronic particles like protons and neutrons. When the age of the Universe passed 1 sec, the temperature became 10^{10} K; the universe became transparent to neutrinos. Neutrinos don't interact with other materials, so they continue as 'Identified remains' of the Great Big Bang. When the age of the Universe was one minute, the process of making nuclei through nuclear fusion became vigorous. Neutrons and protons combined to form first Deuterium and then Tritium [10]. Further, within the first three minutes of the life of the Universe, trillions and trillions of nuclei were produced, and the temperature of the Universe was nearly 10^{9} K. When the age of the Universe was nearly 300,000 years, the temperature reduced to 3000 K. Further, the nuclei captured electrons and formed neutral atoms. When the photon ceased their interactions with free electrons, the Universe became transparent to light for its travel. As a result, on the last scattering surface, a 'Cosmic Background radiation' was formed. After one billion years of its evolution, the temperature of the Universe came down to 20 K. After a few billion years of the Big Bang, our galaxy, the Milky Way was formed. After another 10 billion years, our Sun and Earth were formed. At present, the age of the universe is nearly 14 billion years, and the background temperature is 2.7 K [11].

1.2.2. The Oscillation Theory

Earnest J. Opic put the theory forward regarding the creation of the Universe. According to this theory, there are two possibilities for the evolution of the Universe. Either the Universe should be subject to a change under the flow of time, or it should not be subject to it. From the analysis of the red shift, it is found that the Universe is expanding continuously, and the expansion is continuous for an infinite time; the expansion rate gradually decreases and ultimately disappears [12]. So, in this case, the stars at the end of their lives lose their entire energy and become invisible, diminishing the intensity of galaxies. Moreover, when the Universe is cooled, it loses energy and hence becomes weak and ultimately vanishes. The other alternative is that the expansion rate gradually decreases; after a stage, it contracts in the opposite direction and finally collapses. In this case, when the expansion creases, it contracts into the opposite side, showing a blue shift. Consequently, the stars and galaxies will come together, forming a super-dense super-hot body similar to the primordial one and collapse. This event of collapse is called the Big Crunch. Again by an explosion, the universe will start expanding [13]. In short, the universe behaves like a gigantic oscillating system-Explosion-Expansion-Contraction in the opposite direction- The big crunch-again explosion and so on. The period between two explosions is approximately 25-60 billion years.

1.2.3. The Steady-State Theory

According to this theory Herman Bondi, Thomas Gold, and Fred Hoyle put forward, the Universe has neither a beginning nor an end. This model states that matter in the form of hydrogen atoms is evolved very slowly and continuously and forms galaxies. The process is very slow and hence not observable. The galaxies thus formed get separated and go beyond our observable limit. New galaxies are formed to fill their place so that the average structure of the Universe remains the same always. But we do not have any physical theories supporting matter continuously evolving in the Universe [14].

1.3. Formation of Star

Star formation is the process by which dense regions within molecular gasses in astronomical space, occasionally mentioned as "stellar nurseries" or "star-forming regions," collapse and form stars [15]. Star formation encompasses the study of the giant molecular clouds (GMC) and interstellar medium (ISM) as originators of the star formation progression and the study of protostars and undeveloped stellar entities as its instantaneous produces. It is meticulously associated with planet formation, an alternative subdivision of astronomy. The star formation model, as well as accounting for the development of a single star, must also explain the statistics of dualistic stars and the preliminary mass function. Most stars do not form in isolation but as part of a collection of stars discussed as stellar associations or star clusters [16].

A star is just a huge ball of hot gas heated from the inside by nuclear energy. This enormous energy in their interior is radiated into space and, on a dark night, can be visible to the naked eye. In a sense, the stars are like us. They have a birth, life and death. The infinite space around us is not an absolute vacuum. Instead, even beyond the atmosphere, it is full of gas with a density of one atom or a minute condensed matter of dust per cubic centimeter. In certain regions of space, gigantic layers of clouds of gas and dust are formed with a slightly higher density. These clouds are known as Nebulae. These clouds of atoms and molecules are real cradles of stars [17].

There are two forces that control the process of star formation, namely, a force acting inward due to gravity and a force acting outward due to internal pressure that cools down the cloud. The former tries to contract the cloud, and the latter tries to resist it. The gravitational force contracts the cloud producing a rise in temperature. As the temperature is raised, the speed of motion of molecules increases. This increased speed produces greater kinetic energy and greater pressure at the centre region of the cloud, and that in turn resists the compression of the cloud by the gravitation force. If the compression is to be continued for the formation of a star, the gravitation force should overcome the internal pressure of the cloud due to kinetic energy. In short, for the formation of a star by way of contraction, the cloud must have a mass enough to provide the required gravitation force to overcome the internal pressure due to gas and dust kinetic energy. This minimum mass required for the continuity of star formation is called 'critical mass.' Further, it can be explained another way that when two molecules or atoms collide, their kinetic energy is temporarily converted into the excitation energy in the medium. This makes the electrons in the clouds jump from their permitted orbits. The net energy consequent to the return of the electrons to the permitted orbits results in the emission of energy in the form of photons. This kind of loss or emission of energy cools down the medium; hence the internal pressure that acts outward decreases. This results in the overall contraction of the medium for star formation [18].

Another phenomenon by which stars are formed is the action of solar winds. When striking the interstellar medium, the matter in the solar wind that blows at a speed of a few km/sec from a star compresses it and adds to the chances of star formation.

The massive stars are formed in a much different way, which is a kind of chain reaction. Most of the O and B giant stars are formed within the range of molecular clouds. The heavy stars take one million years to form. Once formed, they produce strong streams of ultraviolet rays, which ionize the medium and increase the temperature forming hydrogen-II regions. With the help of radiation pressure produced by the U-V stream, these H-II regions enter the molecular clouds and compress them with shockwaves [19]. This process accelerates the accretion of matter, which, in turn, helps the formation of stars. The Orion Nebula- a big H-II region over the surface of a molecular complex, is a good example of this kind of star formation.

1.4. Thermonuclear Reactions in Stars

Stars emit an enormous amount of energy and continue to do so for thousands of years. It was suggested that the slow gravitational contraction of a star might release sufficient energy to account for its observed luminosity, but this would limit the life of the sun to 20 million years. So, this rules out the possibility of gravitational contraction providing the main source of the sun's energy [20].

Later, the possibility of “sub atoms” energy sources was considered soon after the discovery of the radioactivity of Uranium and Thorium; it was suggested that they might be a possible source of stellar energy. -lives of these elements seemed to account for the star's long life. However, the abundance of these elements in stellar interiors is much too low to supply enough energy. Moreover, natural radioactive decay rates are unaffected by temperature and pressure variations, so there would be no automated energy release process maintaining the star in equilibrium.

At one time, it was suggested by Jeans that protons and electrons of high energy might collide and annihilate one another converting their masses into energy, but later it was proved that electrons and positrons could do this by radiating γ- rays, but not electrons and protons.

By 1920, many physicists and astronomers were seriously inclined to believe that nuclear fusion reactions between the lightest nuclei were responsible for stellar energy production. In 1927, Atkinson and Houtermans laid down the essential ideas of stellar thermonuclear reactions. They realized that deep in stellar interiors, where the temperature exceeds 10^6 degrees, the matter is completely ionized and that the energy of the nuclei should follow the Maxwell-Boltzmann distribution law [21, 22]. The relatively few nuclei with energies ten or twenty times the most probable value (KT) would eventually penetrate the Coulomb barrier and fuse together, releasing energy in an exoergic reaction that would be very sensitive to temperature.

The major advancement was made in 1938 when the first detailed specific nuclear reactions responsible for Sun's energy were put forward independently by Bethe and Von Weizsakcer. They concluded that two possible processes were responsible for Sun's energy. In each process, the overall effect is to convert hydrogen into helium [23]. The first process involves the interaction of two protons to produce deuterium, the capture of a proton by deuterium to form helium 3, and finally, the interaction of two helium 3 nuclei to form helium 4 and two protons. This is known as the proton-proton chain (p-p). The other possible process is the Carbon Nitrogen cycle, in which Carbon and Nitrogen act as nuclear catalysts in synthesizing helium from hydrogen. The details of these reactions are explained below:

1.4.1. Proton-Proton Cycle

The proton-proton chain reactions on the Sun are:

(a) ${}_1H^1 + {}_1H^1 \rightarrow {}_1H^2 + e^+ + \upsilon + 0.42$ MeV

(b) ${}_1H^2 + {}_1H^1 \rightarrow {}_2He^3 + \gamma + 5.5$ MeV

(c) $_2He^3 + {}_2He^3 \rightarrow 2\ {}_1H^1 + {}_2He^4 + 12.8$ MeV

For reaction (c) to occur, reactions (a) and (b) must occur. When these are written out in detail, and all the reactions are added, we get

$$2_1H^1,\ 2_1H^1 + 2_1H^1 + 2_1H^2 + {}_2He^3 + {}_2He^3 \rightarrow 2_1H^2 + 2e^+ + 2\upsilon + 2_2He^3 + 2\gamma + {}_2He^4 + 2_1H^1 + 24.64 \text{ Mev}$$

So, the net result is

$$4^1{}_1H \rightarrow {}^4{}_2He + 2e^+ + 2\upsilon + 2\gamma + 24.64 \text{ MeV}$$

Since the neutrinos escape, we have left about 24.3 MeV energy after the fusion of four protons to produce one helium nucleus. So, there is a gradual depletion of protons and a building up of the helium concentration [24].

Theoretical aspects indicate that the cross-section of reaction (a) is very low. The very small probability of this reaction occurring during the short collision time of two colliding protons is a consequence of the fact that compound nucleus ^2He must be formed and undergo beta decay to form a deuteron ($_1H^2$) in this time. This is very unlikely as the ^2He system must have the two proton spins antiparallel to satisfy the Pauli Exclusion Principle, and one of the protons must reverse its spin axis when the positron e+ and neutrinos are emitted. The mean lifetime of a proton at the centre of the sun is estimated to be 7 x 10^9 years. Once the deuteron is formed, it reacts after 4 sec with a proton to form $_2He^3$. The final reaction of the p-p chain is slow as the probability of two $_2He^3$ nuclei colliding is very small. Another factor inhibiting the reaction is the greater coulomb repulsion. On average, the lifetime of the He^3 nucleus at the centre of the sun is of the order of 4 x 10^5y [25].

1.4.2. Carbon Nitrogen Cycle

M.A. Bethe is 1939, suggested the following series of reactions are responsible for energy production in the main sequence stars.

(a) $_6C^{12} + {}_1H^1 \rightarrow {}_7N^{13} + \gamma + 1.95$ MeV

(b) $_7N^{13} \rightarrow {}_6C^{13} + e^+ + \upsilon + 2.22$ MeV

(c) $_6C^{13} + {}_1H^1 \rightarrow {}_7N^{14} + \gamma + 7.54$ MeV

(d) $_7N^{14} + {}_1H^1 \rightarrow {}_8O^{15} + \gamma + 7.35$ MeV

(e) ${}_8O^{15} \rightarrow {}_7N^{15} + e^+ + \upsilon + 2.7$ MeV

(f) ${}_7N^{15} + {}_1H^1 \rightarrow {}_6C^{12} + {}_2He^4 + 4.96$ MeV

The net result is ${}^{12}_6C + 4\,{}^1_1H \rightarrow {}^{12}_6C + {}^4_2He + 2e^+ + 2\upsilon + \gamma + 26.72$ MeV

It is the fusion of four protons to produce one He^4 nucleus in the presence of ${}_6C^{12}$, which must be present but is not destroyed in the cycle. So, it acts like catalysts. About 1.7 MeV energy is carried away by the neutrinos, which escape; the net energy release is 25.02 MeV. The average lifetimes of ${}_6C^{12}$, ${}_7N^{13}$, ${}_6C^{13}$, ${}_7N^{14}$, ${}_8O^5$, and ${}_7N^{15}$ nuclei at the centre of the sun before they are destroyed show a range from $2x10^7$y to 2 min [26]. The present evidence is that for the stars with masses between 0.4 to 2.5 solar masses, the main part of energy production is due to the carbon cycle rather than the p-p cycle. For less massive stars, the situation is reversed. So, we can say that whereas the carbon cycle probably goes on in some very bright stars, most stars derive their energy from the p-p cycle [27].

1.4.3. Helium Burning

This is the process where carbon is built up from helium, and the subsequent synthesis of 0^{16} and Ne^{20} by successive α-particle capture occurs. Bethe rejected the possibility of element synthesis in stars because carbon cannot be synthesized from helium. The difficulty is that Be^8 formed by the fusion of two α-particles would break up into two α-particles within 10^{-15}s. Bethe considered the probability of mutual collision of three α-particles leading to the formation of C^{16} very small [28]. However, E. Salpeter showed that under stellar conditions of the density of about 10^5 g/cm^3 and temperature of about 10^8 K, the helium core of the red grant would be converted into carbon [29]. A statistical equilibrium is set up so that the rate of formation of Be^8 is equal to its rate of fusion. Once some Be^8 is produced, C^{12} can be formed by α-capture. The basic reactions of helium burning are:

$${}_2He^4 + {}_2He^4 \leftrightharpoons {}_4Be^8$$

$${}_4Be^8 + {}_2He^4 \leftrightharpoons {}_6C^{12} \rightarrow {}_6C^{12} + \gamma$$

Be^8 (α, γ) C^{12} is a resonant thermonuclear reaction.

Once an appreciable amount of C^{12} has produced an α-particle, it will result in the production of O^{16}, Ne^{20} and Mg^{24}.

1.5. Nucleosynthesis

It is a process that creates new atomic nuclei from nucleons. The first nuclei were formed from Big Bang nucleosynthesis. Then the universe cooled, and these processes ended, leaving the Universe with 75% Hydrogen, 24% of He and traces of other elements. Further heavier elements were created from these elements by several processes. There are different types of nucleosynthesis. The major types are Big Bang nucleosynthesis, stellar nucleosynthesis and explosive or supernova nucleosynthesis.

1.5.1. Big Bang Nucleosynthesis

Big Bang nucleosynthesis occurred within the first three minutes of the beginning of the Universe and is responsible for the abundance of ${}_1H^1$, ${}_1H^2$, ${}_2He^3$, and ${}_2He^4$. Most of the mass of the isotopes in the Universe is thought to have been produced in the Big Bang [30]. The nuclei of these elements, along with some ${}_3Li^7$ and ${}_4Be^7$, are considered to have been formed between 100 to 300s after Big Bang. Because the nucleosynthesis was stopped by expansion and cooling within a few minutes, no elements heavier than beryllium could be formed. The elements formed in this way were in the plasma state and did not cool to the state of neutral atoms until much later [31].

1.5.2. Stellar Nucleosynthesis

The concept was developed by Fred Hoyle. He did not believe in the Big Bang theory, but instead, that hydrogen was continually being created within our universe, and this alternative theory was called a steady-state theory [32]. It is a nuclear process in which new nuclei are produced. It is responsible for the abundance of elements from carbon to iron. It occurs in the stars during stellar evolution. H and He are fused into heavier nuclei at increasingly high temperatures in stars. The most important is carbon which is produced by the triple-alpha process. Carbon causes the release of free neutrons within stars giving rise to the s-process, in which the slow absorption of neutrons converts iron into elements heavier than iron and nickel [33].

To build nuclei of mass number A greater than 60, a source of free neutrons is required. In stars, with burning cores, there is a supply of He^4 and C^{12}. If there is some mixing between hydrogen and helium-burning Zones, ${}^{13}C$ will be produced by the (p, γ) process on C. The abundance of ${}^{13}C$ in the C-N cycle of the hydrogen-burning zone is not likely to be high enough to release sufficient neutrons by the reaction:

$$^{13}{}_{6}C + {}_2He^4 \rightarrow {}_8O^{16} + {}_0n^1 + 2.2\ \text{Mev}$$

Other possible neutron sources are

$$_8^{17}O + _2He^4 \rightarrow _{10}Ne^{20} + _0n^1 + 0.60\text{ MeV}$$

$$_{10}Ne^{20} + _2He^4 \rightarrow _{12}Mg^{24} + _0n^1 + 2.58\text{ MeV}$$

$$_{12}Mg^{25} + _2He^4 \rightarrow Si^{028} + _0n^1 + 2.57\text{ MeV}$$

The Ne^{21} (α, n) Mg^{24} reaction might occur where Ne^{20} mixes with the hydrogen zone of a red giant and produces Ne^{21} by (n, γ) reaction.

1.5.3. Supernova Nucleosynthesis

Supernova nucleosynthesis occurs in the supernova's energetic environment, in which the elements between Silicon and Nickel are synthesized in quasi-equilibrium established during fast fusion. Supernova nucleosynthesis occurs too rapidly for radioactive decay to decrease the number of neutrons, so many abundant isotopes with equal and even numbers of protons and neutrons are synthesized by the silicon quasi-equilibrium process. During this process, the burning of oxygen and silicon fuses nuclei that have an equal number of protons and neutrons that produce nuclides consisting of whole numbers of helium nuclei, up to 15 (^{60}Ni). Such nuclides are stable up to ^{40}Co. So, during Supernova nucleosynthesis, the r-process creates very neutron-rich heavy isotopes, which decay after the event to the first stable isotope, thereby creating neutron-rich stable isotopes of all heavy elements [34, 35].

1.6. Process for Production of Elements

There are different processes for the production of elements that are explained below:

a. **Hydrogen Burning:** This is a process for producing isotopes of C, N, O, F, Ne and Na. It describes the thermonuclear reactions responsible for converting hydrogen into helium by the proton-proton cycle or Carbon Nitrogen cycle.
b. **Helium Burning:** This is a process where carbon is built up from helium and subsequent synthesis of 0^{16}, Ne^{20} by successive and particle capture.
c. **The α-Process:** This is a type of nuclear fusion reaction by which stars convert helium into heavier elements. Some of the reactions are:

$$Ne^{20} + \gamma \rightarrow 0^{16} + H^4 - 4.75\text{ Mev}$$

$$Ne^{20} + He^4 \rightarrow Mg^{24} + \gamma + 9.31\text{ Mev}$$

Two reactions released net energy of 4.56 MeV. Once an appreciable amount of Mg^{24} has been created, further (α, γ) processes will occur.

a. **The e-Process:** If the central temperature exceeds 3 x 10^9 degrees and density is in the range of 10^5 to 10^9 g/cm^3, a large number of nuclear reactions such as (γ, α), (γ, p) (γ, n), (α, n), (p, γ), (n, γ), (p, n) compete and state of statistical equilibrium is reached. The central core nuclei will tend to form the most stable nuclei in the neighborhood of Fe^{56}. The star has a layer structure with different nuclear processes occurring in different regions. The e-process occurs at a fast rate, and an iron group of nuclei may be produced in a few seconds. It is possible that the e-process immediately precedes a Supernova burst.
b. **The s-Process:** To produce elements of Mass number > 60, sources of neutron must occur in stellar evolution. Once neutrons are available, heavy elements can be synthesized as there is no Coulomb barrier inhibiting neutron capture. If conditions are such that neutrons are captured by a given nucleus at a slow rate compared with the half-life of any unstable beta isotope produced, then the nucleus will decay to its daughter, one place higher up in the periodic table before capturing the next neutron. This neutron capture process is known as the s-process.
c. **The r-Process:** This is also known as the rapid neutron capture process. If at some stage of a star, an enormous neutron flux is unleashed, neutrons will be captured so rapidly by certain nuclei that they do not have time to undergo beta decay. This is called the r-process. Many elements in the range $70 < A < 209$ are produced this way.
d. **The p-Process:** This is a process to produce proton-rich nuclei by proton capture (p, γ) or by the equivalent (r, n) process

1.7. Death of Star

Like every living thing on earth, stars are born, mature, reach old age and eventually die. But their life span is measured not in few hundreds or thousands of years but millions and often billions of years. How long a star lives depends mostly on its mass. Small stars with a low mass live the longest, with life spans of tens of billions of years. The most massive stars, however, lead much shorter lives. They may expire after just a few million years.

1.7.1. Death of Sun-like Stars

A small star-like sun stops steadily shining when it runs out of 'Hydrogen fuel' in its core that it has been using since it was born. The core then collapses, releasing energy that makes the outer layers of stars balloon out. The stars expand maybe as much as 50 times to become a 'Red Giant' (Red because its surface is much cooler). Eventually, the core of the giant star shrinks further to become a superdense body called a 'White Dwarf.' Meanwhile, the outer layers of the star have been puffed off into the space, creating one of the most beautiful 'Planetary Nebula.'

1.7.2. Death of Massive Stars

A star with a much greater mass than the sun swells up when it begins to die. Since it is so massive, o it swells up to enormous proportions to become a 'Super Giant.' But it does not remain supergiant for long then its central core collapses in on itself violently, releasing so much energy that it triggers off an explosion that blows the star to bits. The catastrophic explosion is called 'Supernova.' Now, two possibilities arise. If the mass of the star's collapsing core is up to three times the mass of the sun, it becomes a tiny body made up of subatomic particles called neutrons, squashed tightly together, or we call it 'Neutron Star.' Rapidly spinning, it releases pulses of energy like a celestial lighthouse. When the pulses are detected, it is called 'Pulsar.'

If, on the other hand, the mass of the collapsing star is more than three times the mass of the sun, the collapse continues beyond the neutron star stage. The matter in the core is virtually crushed into nothing. All that remains is a region of space dominated by the gravity of the former core matter. The gravity is so great that the region of space will swallow up everything that comes nearby, not even light can escape its clutches, so it is called 'Black Hole.'

1.8. Age of our Galaxy

The age of our galaxy is estimated based on the assumption that the transuranic elements are built by the r-process [36, 37]. Nuclei with mass number 235 + 4n, where n is an integer, will decay by α- particle emission to form ^{235}U, whereas the formation of ^{238}U is restricted. Hoyle estimate the production ratio of ^{235}U to ^{238}U *i.e.*

No of atoms of U^{235} built-in r-process = 1.64

Present-day analysis of uranium ores shows that the atomic abundance ratio of ^{235}U to ^{238}U is 0.0072. Now, half-lives of ^{235}U and ^{238}U are 7.13 10^8 y and 4.51 x 10^9 y, respectively. If all the uranium in our galaxy was the result of a single supernova explosion t_o years ago, t_o can be calculated as:

$$n = n_o \exp\left\{\frac{-0.693 \times t_o}{7.13 \times 10^8}\right\}$$

$$n' = n_o' \exp\left\{\frac{-0.693 \times t_o}{4.51 \times 10^9}\right\}$$

Where n_o and n_o' are the original number of atoms of U^{235} and U^{238}, respectively (t_o years ago). n and n' are the present-day values of ^{235}U and ^{238}U, respectively.

Dividing the equations and putting

$$\frac{n}{n'} = 0.0072 \text{ and } \frac{n_o}{n_o'} = 1.64$$

we have

$$0.0072 = 1.64\, exp\left\{\frac{0.693}{10^9} t_o \left(\frac{1}{4.51} - \frac{1}{0.71}\right)\right\}$$

Therefore,

$$\left\{\frac{0.693}{10^9} t_o \left(\frac{1}{4.51} - \frac{1}{0.71}\right)\right\} = \ln\left(\frac{1.64}{0.0072}\right)$$

So,

$$t_o = 6.6 \text{ x } 10^9 \text{ y}$$

By some other means t_o was found to be 11.5 x 10^9 y, so the age of our galaxy is estimated to be in the range of 6.6 x 10^9 - 11.5x 10^9 years.

REFERENCES

[1] H.E. Suess, and H.C. Urey, "Abundances of the Elements", *Rev. Mod. Phys.*, vol. 28, no. 1, pp. 53-74, 1956.
[http://dx.doi.org/10.1103/RevModPhys.28.53]

[2] J. José, and C Iliadis, "Nuclear astrophysics: the unfinished quest for the origin of the elements", *Rep. Prog. Phys.*, vol. 74, no. 9, p. 096901, 2011.
[http://dx.doi.org/10.1088/0034-4885/74/9/096901]

[3] X.D. Ang, "New Determination of the Astrophysical S Factor SE1 of the C12(α,γ)O16 Reaction (PDF)", *Phys.Rev.Lett.*, vol. 99, no. 5, p. 052502, 2007.

[4] Dan Maoz, *Astrophysics in a Nutshell.* Princeton University Press.p. 272.

[5] Elcio Abdalla, Guillermo Franco Abellán,et al., Shao-Jiang Wang, Richard Watkins, Scott Watson, John K. Webb, Neal Weiner, Amanda Weltman, Samuel J. Witte, Radosław Wojtak, Anil Kumar Yadav, Weiqiang Yang, Gong-Bo Zhao, Miguel Zumalacárregui, "Cosmology intertwined: A review of the particle physics, astrophysics, and cosmology associated with the cosmological tensions and anomalies", *J. High Energy Astrophys.*, 34, 2022, 49-211, ISSN 2214-4048. https://doi.org/10.1016/j.jheap.2022.04.002.

[6] Keeler, E. James, "The Importance of Astrophysical Research and the Relation of Astrophysics to the

Other Physical Sciences", *Astrophys. J.*, vol.6, no.4, pp.271-288, 1987. [http://dx.doi.org/10.1086/140401] [PMID: 17796068]

[7] Hetherington, Norriss S.; McCray, W. Patrick, Weart, Spencer R. (ed.), Spectroscopy and the Birth of Astrophysics, American Institute of Physics, Center for the History of Physics, archived from the original on September 7, 2015, retrieved July 19, 2015.

[8] Francesco Lucchin, Clustering in the Universe, Editor(s): Luciano Pietronero, Erio Tosatti, Fractals in Physics, Elsevier, 1986, Pages 313-318, ISBN 9780444869951. https://doi.org/10.1016/B978-0-4-4-86995-1.50060-3.

[9] C. Stefano, C. Yves, C. H. Eugenio, Wu. Jun, C. Eddy, Electromagnetic Forces and related Mechanical Effects in the Net Plasma Facing Components, Editor(s): B.E. Keen, M. Huguet, R. Hemsworth, Fusion Technology 1990, Elsevier, 1991, Pages 381-385, ISBN 9780444885081. https://doi.org/10.1016/B978-0-444-88508-1.50059-1.

[10] NASA/WMAP Science Team, "Cosmology: The Study of the Universe", *Universe 101: Big Bang Theory,* Washington, D.C, 2011.

[11] Bridge, Mark (Director), "First Second of the Big Bang. How The Universe Works", *Silver Spring, MD. Science Channel,* 2014.

[12] Dorminey Bruce, "The Beginning to the End of the Universe: The mystery of dark energy", Astronomy.com

[13] Hawking, W. Stephen "Particle creation by black holes". Communications in Mathematical Physics. 43 (3): 199–220. Bibcode:1975CMaPh.43.199H. 1975. [http://dx.doi.org/10.1007/BF02345020]

[14] "The Planck Epoch", *The Universe Adventure,* Lawrence Berkeley National Laboratory: Berkeley, CA, 2007.

[15] S.W. Stahler, and F. Palla, *The Formation of Stars.* Wiley-VCH: Weinheim, 2004. [http://dx.doi.org/10.1002/9783527618675]

[16] C.J. Lada, and E.A Lada, "Embedded Clusters in Molecular Clouds", *Annu. Rev. Astron. Astrophys.,* vol. 41, no. 1, pp. 57-115, 2003. [http://dx.doi.org/10.1146/annurev.astro.41.011802.094844]

[17] C. R. O'Dell, "Nebula", *World Book at NASA. World Book, Inc,* 2005.

[18] D. Prialnik, *An Introduction to the Theory of Stellar Structure and Evolution.* Cambridge University Press, 2000, pp. 195-212.

[19] C. Dupraz, and F. Casoli, "The Fate of the Molecular Gas from Mergers to Ellipticals", *Proceedings of the 146th Symposium of the International Astronomical Union.*

[20] E.M. Burbidge, G.R. Burbidge, W.A. Fowler, and F. Hoyle, "Synthesis of the Elements in Stars", *Rev. Mod. Phys.,* vol. 29, no. 4, pp. 547-650, 1957. [http://dx.doi.org/10.1103/RevModPhys.29.547]

[21] S.R. Taylor, "Abundance of chemical elements in the continental crust: a new table", *Geochim. Cosmochim. Acta,* vol. 28, no. 8, pp. 1273-1285, 1964. [http://dx.doi.org/10.1016/0016-7037(64)90129-2]

[22] D.D. Clayton, *Principles of Stellar Evolution and Nucleosynthesis.* University of Chicago Press, 1968.

[23] A.S. Eddington, "The internal constitution of the stars", *Science,* vol. 52, no. 1341, pp. 233-240, 1920. [http://dx.doi.org/10.1126/science.52.1341.233] [PMID: 17747682]

[24] "The Proton–Proton Chain", *Astronomy 162: Stars, Galaxies, and Cosmology,* 2016.

[25] M. Salaris, and S. Cassisi, *Evolution of Stars and Stellar Populations.* John Wiley and Sons, 2005, pp. 119-121. [http://dx.doi.org/10.1002/0470033452]

[26] M. Agostini, K. Altenmüller, S. Appel, V. Atroshchenko, Z. Bagdasarian, D. Basilico, G. Bellini, J. Benziger, R. Biondi, D. Bravo, B. Caccianiga, F. Calaprice, A. Caminata, P. Cavalcante, A. Chepurnov, D. D'Angelo, S. Davini, A. Derbin, A. Di Giacinto, V. Di Marcello, X.F. Ding, A. Di Ludovico, L. Di Noto, I. Drachnev, A. Formozov, D. Franco, C. Galbiati, C. Ghiano, M. Giammarchi, A. Goretti, A.S. Göttel, M. Gromov, D. Guffanti, Aldo Ianni, Andrea Ianni, A. Jany, D. Jeschke, V. Kobychev, G. Korga, S. Kumaran, M. Laubenstein, E. Litvinovich, P. Lombardi, I. Lomskaya, L. Ludhova, G. Lukyanchenko, L. Lukyanchenko, I. Machulin, J. Martyn, E. Meroni, M. Meyer, L. Miramonti, M. Misiaszek, V. Muratova, B. Neumair, M. Nieslony, R. Nugmanov, L. Oberauer, V. Orekhov, F. Ortica, M. Pallavicini, L. Papp, L. Pellicci, Ö. Penek, L. Pietrofaccia, N. Pilipenko, A. Pocar, G. Raikov, M.T. Ranalli, G. Ranucci, A. Razeto, A. Re, M. Redchuk, A. Romani, N. Rossi, S. Schönert, D. Semenov, G. Settanta, M. Skorokhvatov, A. Singhal, O. Smirnov, A. Sotnikov, Y. Suvorov, R. Tartaglia, G. Testera, J. Thurn, E. Unzhakov, F.L. Villante, A. Vishneva, R.B. Vogelaar, F. von Feilitzsch, M. Wojcik, M. Wurm, S. Zavatarelli, K. Zuber, G. Zuzel. The BOREXINO Collaboration "Experimental evidence of neutrinos produced in the CNO fusion cycle in the Sun", *Nature*, 587(7835):577-582. 2020. DOI:10.1038/s41586-020-2934-0.

[27] I.N. Reid, and S.L. Hawley, *"The structure, formation, and evolution of low-mass stars and brown dwarfs – Energy generation". New Light on Dark Stars: Red dwarfs, low-mass stars, brown dwarfs. Springer-Praxis Books in Astrophysics and Astronomy.* 2nd ed. Springer Science & Business Media, 2005, pp. 108-111.

[28] J. Carl, Hansen, D., Steven Kawaler, T. Virginia, Stellar Interiors: Physical Principles, Structure, and Evolution(2 Ed.). Virginia (2004), Springer. pp. 62–5. ISBN 978-0387200897. DOI: 10.1007/978---4419-9110-2

[29] M.A. Seeds, and D.E. Backman, *Foundations of Astronomy.* 12th ed. Cengage Learning, 2012, pp. 249-251.

[30] C. Patrignani, "Big-Bang nucleosynthesis (PDF)", *Chin. Phys. C.,* vol. 40, p. 100001, 2016.

[31] A. Coc, and E. Vangioni, "Primordial nucleosynthesis", *Int. J. Mod. Phys. E,* vol. 26, no. 8, p. 1741002, 2017. [http://dx.doi.org/10.1142/S0218301317410026]

[32] D. D. Clayton, (1983). Principles of Stellar Evolution and Nucleosynthesis (Reprint ed.). Chicago, IL: University of Chicago Press. Chapter 5. ISBN 978-0-226-10952-7.

[33] D.D. Clayton, W.A. Fowler, T.E. Hull, and B.A. Zimmerman, "Neutron capture chains in heavy element synthesis", *Ann. Phys.,* vol. 12, no. 3, pp. 331-408, 1961. [http://dx.doi.org/10.1016/0003-4916(61)90067-7]

[34] S.E. Woosley, W.D. Arnett, and D.D. Clayton, "The Explosive burning of oxygen and silicon", *Astrophys. J. Suppl. Ser.,* vol. 26, pp. 231-312, 1973. [http://dx.doi.org/10.1086/190282]

[35] S. E. Woosley, W. D. Arnett, and D. D. Clayton, "The Explosive Burning of Oxygen and Silicon", *A. J. Supp. Ser.*, vol. 26, p. 231, 1973. DOI:10.1086/190282.

[36] R. Cayrel, V. Hill, T.C. Beers, B. Barbuy, M. Spite, F. Spite, B. Plez, J. Andersen, P. Bonifacio, P. François, P. Molaro, B. Nordström, and F. Primas, "Measurement of stellar age from uranium decay", *Nature,* vol. 409, no. 6821, pp. 691-692, 2001. [http://dx.doi.org/10.1038/35055507] [PMID: 11217852]

[37] J.J. Cowan, C. Sneden, S. Burles, I.I. Ivans, T.C. Beers, J.W. Truran, J.E. Lawler, F. Primas, G.M. Fuller, B. Pfeiffer, and K-L. Kratz, "The Chemical Composition and Age of the Metal-poor Halo Star BD +17o3248", *Astrophys. J.,* vol. 572, no. 2, pp. 861-879, 2002. [http://dx.doi.org/10.1086/340347]

CONCLUSION

This textbook gives an elementary understanding of nuclear and particle physics, offering an overview of theoretical and experimental grounds, providing students with a profound understanding of the ideas about the nucleus, particle detectors, accelerators, radioactivity, and elementary particles. Each chapter provides the fundamental theoretical and experimental knowledge for students to strengthen their concepts regarding nuclear physics. The present form of the textbook is appropriate for undergraduate courses in nuclear and particle physics as well as more innovative courses; the book includes sophisticated and newly constructed figures as well as thoroughly solved equations that create the interest amongst undergraduate students to renovate the content to their course. It could be a vital textbook for students framing their future study or a profession in the field who needs a concrete understanding of nuclear and particle physics together. It provides a concise, thorough, and accessible treatment of the fundamental aspects of nuclear physics. Reorganized figures, resolved equations, rearranged contents and appendices make it easier to use for entire users.

It is an adequate and systematically up–to–date textbook on nuclear and particle physics. Indeed the concepts in this book are unique because it makes important connections to other fields such as elementary particle physics and astrophysics. Moreover, its way of presentation is student-friendly, and it bridges nuclear physics as an essential part of modern physics with a comprehensive scientific and historical context.

The book is intended to focus on the following:

1. Elements of Nuclear Physics is an ultimate textbook for courses at the undergraduate level in nuclear physics. Moreover, it is a significant basis for scientists and those who initiated working with nuclei, particle physicists, astrophysicists and anyone willing to learn more about novel trends in this field.

2. Its importance on phenomenology and discussions of the theory are covered with the suitable examples which clarify and put on the theoretical formulism differentiate this book from all other textbooks available.

3. The text is well organized to deliver the fundamentals of content for students with a small mathematical background that provides more distinguished material in each section.

4. This textbook is competent because it includes the discovery of the neutron and other modern advances, providing undergraduate students with widespread coverage of the elementary models of each topic in particle physics for the first time. Physics emphasizes mathematical precision, making the material handy to students with no earlier knowledge of elementary nuclear physics. The theory and mathematical expressions are linked together in a very sophisticated manner, helping students to understand how key ideas were developed.

Ritesh Kohale, Sanjay J. Dhoble & Vibha Chopra

5. The content on nuclei, mass defect, packing fraction, particle accelerator and detectors, Discovery of Neutron, Nuclear Chain Reaction, Liquid Drop Model and Shell Model of Nucleus, Nuclear fission and fusion, and radioactivity has been completely revised to familiarize the students to what lies beyond. The easily understandable figures and equations with over 50 problems inspire students to apply the theory themselves.

SUBJECT INDEX

A

B

C

Ritesh Kohale, Sanjay J. Dhoble & Vibha Chopra

D

E

O

P

Q

R

S

T

U

V

W

X

www.ingramcontent.com/pod-product-compliance
Lightning Source LLC
LaVergne TN
LVHW070117110826
845147LV00002B/135

* 9 7 8 9 8 1 5 0 4 9 9 2 3 *